Mark Kurlansky

SALT
A World History

VINTAGE BOOKS
London

Published by Vintage 2003

8 10 9 7

First published in Great Britain in 2002 by
Jonathan Cape

Vintage
Random House, 20 Vauxhall Bridge Road,
London SW1V 2SA

www.vintage-books.co.uk

Addresses for companies within The Random House Group Limited
can be found at: www.randomhouse.co.uk/offices.htm

The Random House Group Limited Reg. No. 954009

A CIP catalogue record for this book
is available from the British Library

Illustrations on pages 20, 67, 76, 79, 116, 136, 150, 161,
277, 375, 383, 418, and 420 are by the author. Pages
177–78: Kenneth R. Mackenzie's translation of *Pan Tadeusz*
by Adam Mickiewicz used by permission of
Hippocrene books

Book design by Chris Welch

ISBN 9780099281993

The Random House Group Limited supports The Forest
Stewardship Council (FSC), the leading international forest
certification organisation. All our titles that are printed on
Greenpeace approved FSC certified paper carry the FSC logo.
Our paper procurement policy can be found at
www.rbooks.co.uk/environment.

Mixed Sources
Product group from well-managed
forests and other controlled sources
www.fsc.org Cert no. TT-COC-2139
© 1996 Forest Stewardship Council

Printed and bound in Great Britain by
CPI Bookmarque, Croydon, CR0 4TD

To my parents, Roslyn Solomon and Philip Mendel Kurlansky,
who taught me to love books and music

and

to Talia Feiga, who opened worlds
while she slept in the crook of my arm.

*The real price of every thing, what every thing really costs to
the man who wants to acquire it, is the toil and trouble of
acquiring it.*
—*Adam Smith*, The Wealth of Nations, *1776*

*All our invention and progress seem to result in
endowing material forces with intellectual life, and
in stultifying human life into a material force.*
—*Karl Marx, speech, 1856*

*Dreams are not so different from deeds as some may think.
All the deeds of men are only dreams at first. And
in the end, their deeds dissolve into dreams.*
—*Theodor Herzl*, Old New Land, *1902*

Contents

Contents

Salt

INTRODUCTION

⌒

The Rock

BOUGHT THE rock in Spanish Catalonia, in the rundown hillside mining town of Cardona. An irregular pink trapezoid with
elongated, curved indentations etched on its surface by rain-
drops, it had an odd translucence and appeared to be a cross between rose quartz and soap. The resemblance to soap came from
the fact that it dissolved in water and its edges were worn smooth
like a used soap bar.

I paid too much for it—nearly fifteen dollars. But it was, after
all, despite a rosy blush of magnesium, almost pure salt, a piece
of the famous salt mountain of Cardona. The various families
that had occupied the castle atop the next mountain had garnered centuries of wealth from such rock.

I took it home and kept it on a windowsill. One day it got
rained on, and white salt crystals started appearing on the pink.
My rock was starting to look like salt, which would ruin its mystique. So I rinsed off the crystals with water. Then I spent fifteen

minutes carefully patting the rock dry. By the next day it was sitting in a puddle of brine that had leached out of the rock. The sun hit the puddle of clear water. After a few hours, square white crystals began to appear in the puddle. Solar evaporation was turning brine into salt crystals.

For a while it seemed I had a magical stone that would perpetually produce brine puddles. Yet the rock never seemed to get smaller. Sometimes in dry weather it would appear to completely dry out, but on a humid day, a puddle would again appear under it. I decided I could dry out the rock by baking it in a small toaster oven. Within a half hour white stalactites were drooping from the toaster grill. I left the rock on a steel radiator cover, but the brine threatened to corrode the metal. So I transferred it to a small copper tray. A green crust formed on the bottom, and when I rubbed off the discoloration, I found the copper had been polished.

My rock lived by its own rules. When friends stopped by, I told them the rock was salt, and they would delicately lick a corner and verify that it tasted just like salt.

Those who think a fascination with salt is a bizarre obsession have simply never owned a rock like this.

AMONG THE PEOPLE who have apparently lived with such deprivation was the Welsh Jungian psychologist Ernest Jones, friend of Sigmund Freud and a leading force in bringing psychoanalysis to Britain and the United States. In 1912, Jones published an essay about the human obsession with salt—a fixation that he found irrational and subconsciously sexual. To support his theory, he cited the curious Abyssinian custom of presenting a piece of rock salt to a guest, who would then lick it.

Jones states that "in all ages salt has been invested with a significance far exceeding that inherent in its natural properties, interesting and important as these are. Homer calls it a divine substance, Plato describes it as especially dear to the Gods, and we shall presently note the importance attached to it in religious ceremonies, covenants, and magical charms. That this should have been so in all parts of the world and in all times shows that we are dealing with a general human tendency and not with any local custom, circumstance or notion."

Salt, Jones argued, is often associated with fertility. This notion may have come from the observation that fish, living in the salty sea, have far more offspring than land-based animals. Ships carrying salt tended to be overrun by mice, and for centuries it was believed that mice could reproduce without sex, simply by being in salt.

The Romans, Jones pointed out, called a man in love *salax*, in a salted state, which is the origin of the word *salacious*. In the Pyrenees, bridal couples went to church with salt in their left pockets to guard against impotence. In some parts of France, only the groom carried salt, in others only the bride. In Germany, the bride's shoes were sprinkled with salt.

Jones further built his case: Celibate Egyptian priests abstained from salt because it excited sexual desire; in Borneo, when Dayak tribesmen returned from taking heads, the abstinence from both sex and salt was required; when a Pima killed an Apache, both he and his wife abstained from sex and salt for three weeks. In Behar, India, Nagin women, sacred prostitutes known as "wives of the snake god," periodically abstained from salt and went begging. Half their proceeds were given to the priests and half to buying salt and sweetmeats for the villagers.

An 1157 Paris engraving titled Women Salting Their Husbands *demonstrated how to make your man more virile. The last line of an accompanying poem reads, "With this salting, front and back, At last strong natures they will not lack."* Bibliothèque Nationale

Jones bolstered his argument by turning to Freud, who eight years earlier had asserted in *Zur Psychopathie des Altagslebens*, On the Psychopathology of Daily Life, that superstitions were often the result of attaching great significance to an insignificant object or phenomenon because it was unconsciously associated with something else of great importance.

Would not all this attention to salt be inexplicable, Jones's argument goes, unless we were really thinking of more important things—things worthy of an obsession? Jones concludes, "There is every reason to think that the primitive mind equated the idea of salt, not only with that of semen, but also with the essential constituent of urine."

JONES WAS WRITING in an age with a thirst for scientific explanations. And it is true that semen and urine—along with blood, tears, sweat, and almost every part of the human body—contain salt, which is a necessary component in the functioning of cells. Without both water and salt, cells could not get nourishment and would die of dehydration.

But perhaps a better explanation for the human obsession with this common compound is the one offered a few years later, in the 1920s, by the Diamond Crystal Salt Company of St. Clair, Michigan, in a booklet, "One Hundred and One Uses for Diamond Crystal Salt." This list of uses included keeping the colors bright on boiled vegetables; making ice cream freeze; whipping cream rapidly; getting more heat out of boiled water; removing rust; cleaning bamboo furniture; sealing cracks; stiffening white organdy; removing spots on clothes; putting out grease fires; making candles dripless; keeping cut flowers fresh; killing poison ivy; and treating dyspepsia, sprains, sore throats, and earaches.

Far more than 101 uses for salt are well known. The figure often cited by the modern salt industry is 14,000, including the manufacturing of pharmaceuticals, the melting of ice from winter roads, fertilizing agricultural fields, making soap, softening water, and dying textiles.

Salt is a chemical term for a substance produced by the reaction of an acid with a base. When sodium, an unstable metal that can suddenly burst into flame, reacts with a deadly poisonous gas known as chlorine, it becomes the staple food sodium chloride, NaCl, from the only family of rocks eaten by humans. There are many salts, and a number of them are edible and often found together. The one we most like to eat is sodium chloride, which has the taste that we call salty. Other salts contribute unwelcome bitter or sour tastes, though they may also be of value to the human diet. Baby formula contains three salts: magnesium chloride, potassium chloride, and sodium chloride.

Chloride is essential for digestion and in respiration. Without sodium, which the body cannot manufacture, the body would be unable to transport nutrients or oxygen, transmit nerve impulses, or move muscles, including the heart. An adult human being contains about 250 grams of salt, which would fill three or four saltshakers, but is constantly losing it through bodily functions. It is essential to replace this lost salt.

A French folktale relates the story of a princess who declares to her father, "I love you like salt," and he, angered by the slight, banishes her from the kingdom. Only later when he is denied salt does he realize its value and therefore the depth of his daughter's love. Salt is so common, so easy to obtain, and so inexpensive that we have forgotten that from the beginning of civilization until about 100 years ago, salt was one of the most sought-after commodities in human history.

SALT PRESERVES. UNTIL modern times it provided the principal way to preserve food. Egyptians used salt to make mummies. This ability to preserve, to protect against decay, as well as to sustain life, has given salt a broad metaphorical importance—what Freud might have considered an irrational attachment to salt, a seemingly trivial object, because, in our unconscious, we associate it with longevity and permanence, which are of boundless significance.

Salt was to the ancient Hebrews, and still is to modern Jews, the symbol of the eternal nature of God's covenant with Israel. In the Torah, the Book of Numbers, is written, "It is a covenant of salt forever, before the Lord," and later in Chronicles, "The Lord God of Israel gave the kingdom over Israel to David forever, even to him, and to his sons, by a covenant of salt."

On Friday nights Jews dip the Sabbath bread in salt. In Judaism, bread is a symbol of food, which is a gift from God, and dipping the bread in salt preserves it—keeps the agreement between God and his people.

Loyalty and friendship are sealed with salt because its essence does not change. Even dissolved into liquid, salt can be evaporated back into square crystals. In both Islam and Judaism, salt seals a bargain because it is immutable. Indian troops pledged their loyalty to the British with salt. Ancient Egyptians, Greeks, and Romans included salt in sacrifices and offerings, and they invoked gods with salt and water, which is thought to be the origin of Christian holy water.

In Christianity, salt is associated not only with longevity and permanence but, by extension, with truth and wisdom. The Catholic Church dispenses not only holy water but holy salt, *Sal Sapientia*, the Salt of Wisdom.

Bread and salt, a blessing and its preservation, are often associated. Bringing bread and salt to a new home is a Jewish tradition dating back to the Middle Ages. The British dispensed with the bread but for centuries carried salt to a new home. In 1789, when Robert Burns moved to a new house in Ellisland, he was escorted there by a procession of relatives carrying a bowl of salt. The city of Hamburg, Germany, symbolically renews its blessings once a year by carrying through the streets a chocolate-covered bread and a marzipan saltcellar filled with sugar. In Welsh tradition, a plate was put on the coffin with bread and salt, and a local professional sin eater arrived to eat the salt.

Because salt prevents decay, it protects from harm. In the early Middle Ages, farmers in northern Europe learned to save their grain harvest from a devastating fungal infection called ergot, poisonous to humans and livestock, by soaking the grain in salt brine. So it is not surprising that Anglo-Saxon farmers included salt in the magic ingredients placed in a hole in the plow as they invoked the name of the earth goddess and chanted for "bright crops, broad barley, white wheat, shining millet . . ."

Evil spirits detest salt. In traditional Japanese theater, salt was sprinkled on the stage before each performance to protect the actors from evil spirits. In Haiti, the only way to break the spell and bring a zombie back to life is with salt. In parts of Africa and the Caribbean, it is believed that evil spirits are disguised as women who shed their skin at night and travel in the dark as balls of fire. To destroy these spirits their skin must be found and salted so that they cannot return to it in the morning. In Afro-Caribbean culture, salt's ability to break spells is not limited to evil spirits. Salt is not eaten at ritual meals because it will keep all the spirits away.

Both Jews and Muslims believe that salt protects against the evil eye. The Book of Ezekial mentions rubbing newborn infants with salt to protect them from evil. The practice in Europe of protecting newborns either by putting salt on their tongues or by submerging them in saltwater is thought to predate Christian baptism. In France, until the practice was abolished in 1408, children were salted until they were baptized. In parts of Europe, especially Holland, the practice was modified to placing salt in the cradle with the child.

Salt is a potent and sometimes dangerous substance that has to be handled with care. Medieval European etiquette paid a great deal of attention to how salt was touched at the table—with the tip of a knife and never by hand. The most authoritative book of Jewish law, the *Shulchan Arukh*, The Prepared Table, written in the sixteenth century, it is explained that salt can only safely be handled with the middle two fingers. If a man uses his thumb in serving salt, his children will die, his little finger will cause poverty, and use of the index finger will cause him to become a murderer.

MODERN SCIENTISTS ARGUE about how much salt an adult needs to be healthy. Estimates range from two-thirds of a pound to more than sixteen pounds each year. People who live in hot weather, especially if they do physical labor, need more salt because they must replace the salt that is lost in sweating. This is why West Indian slaves were fed salted food. But if they do not sweat excessively, people who eat red meat appear to derive from it all the salt they need. The Masai, nomadic cattle herders in East Africa, meet their salt needs by bleeding livestock and drinking the blood. But vegetable diets, rich in potassium, offer little sodium chloride. Wherever records exist of humans in different stages of development, as in seventeenth- and eighteenth-century North America, it is generally found that hunter tribes neither made nor traded for salt but agricultural tribes did. On every continent, once human beings began cultivating crops, they began looking for salt to add to their diet. How they learned of this need is a mystery. A victim of starvation experiences hunger, and so the need for food is apparent. Salt deficiency causes headaches and weakness, then light-headedness, then nausea. If deprived long enough, the victim will die. But at no time in this process is a craving for salt experienced. However, most people choose to eat far more salt than they need, and perhaps this urge—the simple fact that we like the taste of salt—is a natural defense.

The other development that created a need for salt was the move to raise animals for meat rather than kill wild ones. Animals also need salt. Wild carnivores, like humans, can meet their salt needs by eating meat. Wild herbivores forage for it, and one of the earliest ways humans searched for salt was to follow animal trails. Eventually they all lead to a salt lick or a brine spring or some other source of salt. But domesticated animals need to be fed salt. A horse can require five times the salt intake of a human, and a cow needs as much as ten times the amount of salt a human requires.

Attempts to domesticate animals may have occurred before the end of the Ice Age, and even then humans understood that animals needed salt. Reindeer were observed going to encampments where human urine provided a source of salt. People learned that if salt was provided, the reindeer would come to them and eventually be tamed. But though these animals became a source of food, they never became truly domesticated animals.

Around 11,000 B.C., the Ice Age ended, and vast sheets of ice that covered much of the known world, including what is today New York and Paris, began to shrink and slowly vanish. At about this time the Asiatic wolf, a fierce predator that despite its small size would eat a human if it had the opportunity, came under human control because its friendly young cubs could be fed and trained. A dangerous adversary was turned into a dedicated helper—the dog.

As glaciers melted, huge fields of wild grain appeared. Humans, but also wild sheep and goats, fed on these fields. The initial human reaction was probably to kill these animals that threatened their food supply. But tribes living near such fields soon realized that sheep and goats could become a dependable food source if they could control them. Their dogs could even help in this work. By 8900 B.C., sheep were domesticated in Iraq, though they may have been domesticated in other places even earlier.

Around 8000 B.C., women in the Near East began planting seeds of wild grains in cleared fields. This is usually thought to have been the beginning of agriculture. But in 1970, a University of Hawaii expedition to Burma, now known as Myanmar, reported finding in a place called Spirit Cave the remnants of what seemed to be cultivated vegetables—peas, water chestnuts, and cucumbers—carbon-dated to the year 9750 B.C.

Pigs came later, about 7000 B.C., because they would not simply graze on grass, and it took time to see the benefit of keeping animals for whom food had to be gathered. It was not until about 6000 B.C. in Turkey or the Balkans that people successfully carried out the daunting task of domesticating the large, fast, and powerful aurochs. Through controlling their diet, castrating males, and corralling the animals into constricted spaces, people eventually turned the wild aurochs into cattle. Cattle became a mainstay food, consuming huge quantities of both grain and salt. The aurochs, fast-footed and ferocious, were hunted into extinction by the mid–seventeenth century.

Where people ate a diet consisting largely of grains and vegetables, supplemented by the meat of slaughtered domestic farm animals, procuring salt became a necessity of life, giving it great symbolic importance and economic value. Salt became one of the first international commodities of trade; its production was one of the first industries and, inevitably, the first state monopoly.

⌒

THE SEARCH FOR salt has challenged engineers for millennia and created some of the most bizarre, along with some of the most ingenious, machines. A number of the greatest public works ever conceived were motivated by the need to move salt. Salt has been in the forefront of the development of both chemistry and geology. Trade routes that have remained major thoroughfares were established, alliances built, empires secured, and revolutions provoked—all for something that fills the ocean, bubbles up from springs, forms crusts in lake beds, and thickly veins a large part of the earth's rock fairly close to the surface.

Almost no place on earth is without salt. But this was not clear until revealed by modern geology, and so for all of history until the twentieth century, salt was desperately searched for, traded for, and fought over. For millennia, salt represented wealth. Caribbean salt merchants stockpiled it in the basements of their homes. The Chinese, the Romans, the French, the Venetians, the Hapsburgs, and numerous other governments taxed it to raise money for wars. Soldiers and sometimes workers were paid in salt. It was often used as money.

In his 1776 treatise on capitalism, *The Wealth of Nations*, Adam Smith pointed out that almost anything of value could be used for money. He cited as examples tobacco, sugar, dried cod, and cattle and stated that "salt is said to be a common instrument of commerce and exchanges in Abyssinia." But he offered the opinion that the best currency was made of metal because it was physically durable, even if its value was as ephemeral as other commodities.

Today, thousands of years of coveting, fighting over, hoarding, taxing, and searching for salt appear picturesque and slightly foolish. The seventeenth-century British leaders who spoke with urgency about the dangerous national dependence on French sea salt seem somehow more comic than contemporary leaders concerned with a dependence on foreign oil. In every age, people are certain that only the things they have deemed valuable have true value.

The search for love and the search for wealth are always the two best stories. But while a love story is timeless, the story of a quest for wealth, given enough time, will always seem like the vain pursuit of a mirage.

PART ONE

A Discourse on Salt, Cadavers, and Pungent Sauces

A country is never as poor as when it seems filled with riches.
—Laozi quoted in the Yan tie lun,
A Discourse on Salt and Iron, 81 B.C.

CHAPTER ONE

A Mandate of Salt

ONCE I STOOD on the bank of a rice paddy in rural Sichuan Province, and a lean and aging Chinese peasant, wearing a faded forty-year-old blue jacket issued by the Mao government in the early years of the Revolution, stood knee deep in water and apropos of absolutely nothing shouted defiantly at me, "We Chinese invented many things!"

The Chinese are proud of their inventions. All Chinese leaders, including Mao Zedong, sooner or later give a speech listing the many Chinese firsts. Though rural China these days seems in need of a new round of inventions, it is irrefutably true that the Chinese originated many of the pivotal creations of history, including papermaking, printing, gunpowder, and the compass.

China is the oldest literate society still in existence, and its 4,000 years of written history begin as a history of inventions. It is no longer clear when legends were made into men and when living historic figures were turned into legends. Chinese history starts in the same manner as Old Testament history. In the Book of Genesis, first come the legends, the story of the Creation, mythical figures such as Adam and Eve and Noah, generations of people who may or may not have lived, and gradually the generations are followed to Abraham, the beginning of documented Hebrew history.

In Chinese history, first was Pangu, the creator, who made humans from parasites on his body. He died but was followed by wise rulers, who invented the things that made China the first civilization. Fuxi was first to domesticate animals. Apparently an enthusiast for domesticity, he is also credited with inventing marriage. Next came Shennong, who invented medicine, agriculture, and trade. He is credited with the plow and the hoe. Then came Huangdi, the Yellow Emperor, who invented writing, the bow and arrow, the cart, and ceramics. Several centuries after Huangdi came Emperor Yao, a wise ruler who passed over his unqualified son and named a modest sage, Shun, his successor. Shun chose his minister, Yu, to succeed him. In 2205 B.C., according to tradition, Yu founded the Xia dynasty, and this dynasty, which lasted until 1766 B.C., enters into documented history.

CHINESE SALT HISTORY begins with the mythical Huangdi, who invented writing, weaponry, and transportation. According to the legends, he also had the distinction of presiding over the first war ever fought over salt.

One of the earliest verifiable saltworks in prehistoric China was in the northern province of Shanxi. In this arid region of dry yellow earth and desert mountains is a lake of salty water, Lake Yuncheng. This area was known for constant warfare, and all of the wars were over control of the lake. Chinese historians are certain that by 6000 B.C., each year, when the lake's waters evaporated in the summer sun, people harvested the square crystals on the surface of the water, a system the Chinese referred to as "dragging and gathering." Human bones found around the lake have been dated much earlier, and some historians speculate that these inhabitants may also have gathered salt from the lake.

The earliest written record of salt production in China dates to around 800 B.C. and tells of production and trade of sea salt a millennium before, during the Xia dynasty. It is not known if the techniques described in this account were actually used during the Xia dynasty, but they were considered old ways by the time of this account, which describes putting ocean water in clay vessels and boiling it until reduced to pots of salt crystals. This was the technique that was spread through southern Europe by the Roman Empire, 1,000 years after the Chinese account was written.

About 1000 B.C., iron first came into use in China, though the first evidence of it being used in salt making is not until 450 B.C. by a man named Yi Dun. According to a passage written in 129 B.C., "Yi Dun rose to prominence by producing salt in pans." Yi Dun is believed to have made salt by boiling brine in iron pans, an innovation which would become one of the leading techniques for salt making for the next 2,000 years. The legend says that he worked with an ironmaster named Guo Zong and was also friendly with an enterprising wealthy bureaucrat named Fan Li. Fan Li is credited with inventing fish farming, which for centuries after was associated with salt-producing areas. The Chinese, like later Europeans, saw that salt and fish were partners. Many Chinese, including Mencius, the famous Confucian thinker who lived from 372 to 289 B.C., were said to have worked selling both fish and salt.

⌒

THROUGHOUT THE LONG history of China, sprinkling salt directly on food has been a rarity. Usually it has been added during cooking by means of various condiments—salt-based sauces and pastes. The usual explanation is that salt was expensive and it was stretched by these condiments. A recurring idea throughout the ancient world from the Mediterranean to Southeast Asia, fish fermented in salt was one of the most popular salt condiments in ancient China. It was called *jiang*. But in China soybeans were added to ferment with the fish, and in time the fish was dropped altogether from the recipe and jiang became *jiangyou*, or, as it is called in the West, soy sauce.

Soy is a legume that produces beans, two or three in a two-inch-long furry pod. The beans can be yellow, green, brown, purple, black, or spotted, and Chinese cooking makes a great distinction among these varieties. Jiangyou is made from yellow beans, but other types are also fermented with salt to produce different pastes and condiments. In China, the earliest written mention of soy is in the sixth century B.C., describing the plant as a 700-year-old crop from the north. Soy was brought to Japan from China in the sixth century A.D. by Chinese Buddhist missionaries. Both the religion and the bean were successfully implanted. But the Japanese did not make soy sauce until the tenth century. Once they did learn, they called it *shoyu* and industrialized it and sold it around the world.

Though *jiangyou* and *shoyu* are pronounced very differently and appear to be very different words in Western writing, the two words are written with the same character in Japanese and Chinese. Mao's 1950s literacy campaign simplified the language to some 40,000 characters, but a pre-Mao character for the soy plant, *su*, depicts little roots at the bottom which revive the soil.

Soy puts nutrients back into the soil and can restore fields that have been exhausted by other crops. The bean is so nutritious that a person could be sustained for a considerable period on nothing but water, soy, and salt.

THE PROCESS BY which the Chinese, and later the Japanese, fermented beans in earthen pots is today known as lactic acid fermentation, or, in more common jargon, pickling. Optimum lactic fermentation takes place between sixty-four and seventy-one degrees Fahrenheit, which in most of the world is an easily achieved environment.

As vegetables begin to rot, the sugars break down and produce lactic acid, which serves as a preservative. Theoretically, pickling can be accomplished without salt, but the carbohydrates and proteins in the vegetables tend to putrefy too quickly to be saved by the emerging lactic acid. Without salt, yeast forms, and the fermentation process leads to alcohol rather than pickles.

Between .8 and 1.5 percent of the vegetable's weight in salt holds off the rotting process until the lactic acid can take over. Excluding oxygen, either by sealing the jar or, more usually, by weighting the vegetables so that they remain immersed in liquid, is necessary for successful lactic fermentation.

The ancient Chinese pickled in earthen jars, which caused a white film called kahm yeast, harmless but unpleasant tasting, to form on the top. Every two weeks the cloth, board, and stone weighting the vegetables had to be washed or even boiled to remove the film. This added work is why pickling in earthen jars has not remained popular.

In Sichuan, pickled vegetables are still a staple. They are served with rice, which is never salted. The salty vegetables contrast pleasantly with the blandness of the warm but unseasoned rice gruel that is a common breakfast food. In effect, the pickles are salting the rice.

South of the Sichuan capital of Chengdu, lies Zigong, a hilly provincial salt town that grew into a city because of its preponderance of brine wells. The crowded, narrow, downhill open-air market in the center of town continues to sell salt and special pickling jars for the two local specialties, *paocai* and *zhacai*. A woman at the market who sold the glass pickling jars offered this recipe for paocai:

> Fill the jar two-thirds with brine. Add whatever vegetables you like and whatever spice you like, cover, and the vegetables are ready in two days.

The spices added are usually hot red Sichuan peppers or ginger, a perennial herb of Indian origin, known to the Chinese since ancient times. The red pepper, today a central ingredient of Sichuan cooking, did not arrive until the sixteenth century, carried to Europe by Columbus, to India by the Portuguese, and to China by either the Indians, Portuguese, Andalusians, or Basques.

Paocai that is eaten in two days is obviously more about flavor than preserving. After two days the vegetables are still very crisp, and the salt maintains, even brightens, the color. Zhacai is made with salt instead of brine, alternating layers of vegetables with layers of salt crystals. In time a brine is formed from the juices the salt pulls out of the vegetables. When a peasant has a baby girl, the family puts up a vegetable every year and gives the jars to her when she's married. This shows how long zhacai is kept before eating. The original medieval idea was to marry her after twelve or fifteen jars. Today it usually takes a few more vegetables.

The Chinese also solved the delicate problem of transporting eggs by preserving them in salt. They soaked the eggs, and still do, in brine for more than a month, or they soak them for a shorter time and encase them in salted mud and straw. The resulting egg, of a hard-boiled consistency with a bright orange yolk, will neither break nor spoil if properly handled. A more complicated technique, involving salt, ash, lye, and tea, produces the "1,000-year-old egg." Typical of the Chinese love of poetic hyperbole, 1,000-year-old eggs take about 100 days to make, and will keep for another 100 days, though the yolk is then a bit green and the smell is strong.

IN 250 B.C., the time of the Punic Wars in the Mediterranean, the governor of Shu, today the province of Sichuan, was a man named Li Bing. The governor was one of the greatest hydraulic engineering geniuses of all time.

The coincidence of hydraulic engineering skills and political leadership does not seem strange when it is remembered that water management was one of the critical issues in developing China, a land of droughts and floods.

The Yellow River, named for the yellowish silt it rushes through northern China, was known as "the father of floods." It and the Yangtze are the two great rivers of Chinese history, both originating in the Tibetan plateau and winding toward the sea on the east coast of China. The Yellow runs through arid northern regions and tends to silt up, raising the riverbed, which causes flooding unless dikes are built up around its banks. The Yangtze is a wider river with many navigable tributaries. It flows through the green and rainy center of China, bisecting the world's third largest country, from the Tibetan mountains to Shanghai on the East China Sea.

The rule of the wise Emperor Yao is said to have been a golden age of ancient China, and one reason for this was that Emperor Yao had tamed nature by introducing the concept of flood control.

Li Bing has taken on some of the mythic dimensions of Yao, a god who conquered floods and tamed nature. But unlike the mythical Emperor Yao, Li Bing's existence is well documented. His most extraordinary accomplishment was the building of the first dam, which still functions in modernized form. At a place called Dujiangyan, he divided the Minjiang River, a tributary of the Yangtze. The diverted water goes into a series of spillways and channels that can be opened to irrigate in times of droughts and closed in times of flooding. He had three stone figures of men placed in the water as gauges. If their feet were visible, this signaled drought conditions and the dam's gates were opened to let in water. If their shoulders were submerged, floodwaters had risen too high and the dam's gates were closed.

Because of the Dujiangyan dam system, the plains of eastern Sichuan became an affluent agricultural center of China. Ancient

In 1974 this statue was found in a river in Sichuan during work on an irrigation system. Inscriptions on the sleeves tell of three statues built "to perpetually guard the waters." On the front of the statue the inscription reads "The late governor of Shu, Li Bing." Ann Paludan

records called the area "Land of Abundance." With the dam still operating, the Sichuan plains remain an agricultural center today.

In 1974, two water gauges, carved in A.D. 168, were found in the riverbed by the site of Li Bing's dam. They seem to have been replacements for the original water gauge statues. One of them is the oldest Chinese stone figure ever found of an identifiable individual. It is a statue of Li Bing. The original gauges he had used depicted gods of flood control. Four centuries after his death, he was considered to be one of these gods.

Li Bing made a very simple but pivotal discovery. By his time, Sichuan had long been a salt-producing area. Salt is known to have been made in Sichuan as early as 3000 B.C. But it was Li Bing who found that the natural brine, from which the salt was made, did not originate in the pools where it was found but seeped up from underground. In 252 B.C., he ordered the drilling of the world's first brine wells.

These first wells had wide mouths, more like an open pit, though some went deeper than 300 feet. As the Chinese learned how to drill, the shafts got narrower and the wells deeper.

But sometimes the people who dug the wells would inexplicably become weak, get sick, lie down, and die. Occasionally, a tremendous explosion would kill an entire crew or flames spit out from the bore holes. Gradually, the salt workers and their communities realized that an evil spirit from some underworld was rising up through the holes they were digging. By 68 B.C., two wells, one in Sichuan and one in neighboring Shaanxi, became infamous as sites where the evil spirit emerged. Once a year the governors of the respective provinces would visit these wells and make offerings.

By A.D. 100, the well workers, understanding that the disturbances were caused by an invisible substance, found the holes where it came out of the ground, lit them, and started placing pots close by. They could cook with it. Soon they learned to insulate bamboo tubes with mud and brine and pipe the invisible force to boiling houses. These boiling houses were open sheds where pots of brine cooked until the water evaporated and left salt crystals. By A.D. 200, the boiling houses had iron pots heated by gas flames. This is the first known use of natural gas in the world.

Salt makers learned to drill and shore up a narrow shaft, which allowed them to go deeper. They extracted the brine by means of a long bamboo tube which fit down the shaft. At the bottom of the tube was a leather valve. The weight of the water would force the valve shut while the long tube was hauled out. Then the tube was suspended over a tank, where a poke from a stick would open the valve and release the brine into the tank. The tank was connected to bamboo piping that led to the boiling house. Other bamboo pipes, planted just below the wellhead to capture escaping gas, also went to the boiling house.

Bamboo piping, which was probably first made in Sichuan, is salt resistant, and the salt kills algae and microbes that would cause rot. The joints were sealed either with mud or with a mixture of tung oil and lime. From the piping at Sichuan brine works, Chinese throughout the country learned to build irrigation and plumbing systems. Farms, villages, and even houses were built with bamboo plumbing. By the Middle Ages, the time of the Norman conquest of England, Su Dongpo, a bureaucrat born in Sichuan, was building sophisticated bamboo urban plumbing. Large bamboo water mains were installed in Hangzhou in 1089 and in Canton in 1096. Holes and ventilators were installed for dealing with both blockage and air pockets.

Salt producers spread out bamboo piping over the countryside with seeming chaos like the web of a monster spider. The pipes were laid over the landscape to use gravity wherever possible, rising and falling like a roller coaster, with loops to create long downhill runs.

In the mid-eleventh century, while King Harold was unsuccessfully defending England from the Normans, the salt producers of Sichuan were developing percussion drilling, the most advanced drilling technique in the world for the next seven or eight centuries.

A hole about four inches in diameter was dug by dropping a heavy eight-foot rod with a sharp iron bit, guided through a bamboo tube so that it kept pounding the same spot. The worker stood on a wooden lever, his weight counterbalancing the eight-foot rod on the other end. He rode the lever up and down, seesaw-like, causing the bit to drop over and over again. After three

Ancient bamboo piping carrying brine through the Sichuan countryside outside Zigong, circa 1915. Zigong Salt History Museum

to five years, a well several hundred feet deep would strike brine.

In 1066, Harold was killed at Hastings by an arrow, the weapon the Chinese believe was invented in prehistory by Huangdi. At the time of Harold's death, the Chinese were using gunpowder, which was one of the first major industrial applications for salt. The Chinese had found that mixing potassium nitrate, a salt otherwise known as saltpeter, with sulfur and carbon created a powder that when ignited expanded to gas so quickly it produced an explosion. In the twelfth century, when European Crusaders were failing to wrest Jerusalem from the infidel Arabs, the Arabs were beginning to learn of the secret Chinese powder.

⌒

LI BING HAD lived during one of the most important crossroads in Chinese history. Centuries of consolidation among warring

The character yan drawn by Beijing salt history professor Guo, in the zhuangzi style of calligraphy that was used until about 200 B.C. Guo Zhenzhong

states had at last produced a unified China. The unified state was the culmination of centuries of intellectual debate about the nature of government and the rights of rulers. At the center of that debate was salt.

Chinese governments for centuries had seen salt as a source of state revenue. Texts have been found in China mentioning a salt tax in the twentieth century B.C. The ancient character for salt, *yan,* is a pictograph in three parts. The lower part shows tools, the upper left is an imperial official, and the upper right is brine. So the very character by which the word *salt* was written depicted the state's control of its manufacture.

A substance needed by all humans for good health, even survival, would make a good tax generator. Everyone had to buy it, and so everyone would support the state through salt taxes.

The debate about the salt tax had its roots in Confucius, who lived from 551 to 479 B.C. In Confucius's time the rulers of various Chinese states assembled what would today be called think tanks, in which selected thinkers advised the ruler and debated among themselves. Confucius was one of these intellectual advisers. Considered China's first philosopher of morality, he was disturbed by human foibles and wanted to raise the standard of human behavior. He taught that treating one's fellow human beings well was as important as respecting the Gods, and he emphasized the importance of respecting parents.

Confucius's students and their students built the system of thought known as Confucianism. Mencius, a student of Confucius's grandson, passed teachings down in a book called the *Mencius*. Confucius's ideas were also recorded in a book called *The Analects*, which is the basis of much Chinese thought and the source of many Chinese proverbs.

During the two and a half centuries between Confucius and Li Bing, China was a grouping of numerous small states constantly at war. Rulers fell, and their kingdoms were swallowed up by more powerful ones, which would then struggle with other surviving states. Mencius traveled in China explaining to rulers that they stayed in power by a "mandate from heaven" based on moral principles, and that if they were not wise and moral leaders, the gods would take away their mandate and they would fall from power.

But another philosophy, known as legalism, also emerged. The legalists insisted that earthly institutions effectively wielding power were what guaranteed a state's survival. One of the leading legalists was a man named Shang, who advised the Qin (pronounced CHIN) state. Shang said that respect for elders and tradition should not interfere with reforming, clearing out inefficient institutions and replacing them with more effective and pragmatic programs. Legalists struggled to eliminate aristocracy, thereby giving the state the ability to reward and promote based on achievement.

The legalist faction had a new idea about salt. The first written text on a Chinese salt administration is the *Guanzi*, which contains what is supposed to be the economic advice of a minister who lived from 685 to 643 B.C. to the ruler of the state of Qi. Historians agree that the *Guanzi* was actually written around 300

B.C., when only seven states still remained and the eastern state of Qi, much under the influence of legalism, was in a survival struggle, which it would eventually lose, with the western state of Qin.

Among the ideas offered by the minister was fixing the price of salt at a higher level than the purchase price so that the state could import the salt and sell it at a profit. "We can thus take revenues from what other states produce." The adviser goes on to point out that in some non-salt-producing areas people are ill from the lack of it and in their desperation would be willing to pay still higher prices. The conclusion of the *Guanzi* is that "salt has the singularly important power to maintain the basic economy of our state."

By 221, Qin defeated its last rivals, and its ruler became the first emperor of united China. China would continue to be ruled by such emperors until 1911.

The proposals in the *Guanzi,* which became Qi policy, now became the policy of the Qin and the emperor of China. The Qin dynasty was marked by the legalists' tendency for huge public works and harsh laws. A price-fixing monopoly on salt and iron kept prices for both commodities excessively high. It is the first known instance in history of a state-controlled monopoly of a vital commodity. The salt revenues were used to build not only armies but defensive structures including the Great Wall, designed to keep Huns and other mounted nomadic invaders from the north out of China. But the harsh first dynasty lasted less than fifteen years.

The Han dynasty that replaced it in 207 B.C., ended the unpopular monopolies to demonstrate better, wiser government. But in 120 B.C., expeditions were still being mounted to drive back the Huns, and the treasury was drained to pay for the wars with "barbarians" in the north. The Han emperor hired a salt maker and ironmaster to research the possibility of resurrecting the salt and iron monopolies. Four years later, he put both monopolies back into place.

China at the time was probably the most advanced civilization on earth at what was a high point of territorial expansion, economic prosperity, and trade. The Chinese world had expanded much farther than that of the Romans. Rome had an empire by conquest, was at the zenith of its power as well, but was menaced by the Gauls and Germanic tribes and even more threatened by internal civil wars. The Chinese had first learned of the Roman Empire in 139 B.C., when the emperor Wudi had sent an envoy, Zhang Qian, past the deserts to seek allies to the west. Zhang Qian traveled for twelve years to what is now Turkistan and back and reported on the astounding discovery that there was a fairly advanced civilization to the west. In 104 B.C. and 102 B.C., Chinese armies reached the area, a former Greek kingdom called Sogdiana with its capital in Samarkand, where they met and defeated a force partly composed of captive Roman soldiers.

In China the salt and iron monopolies, whose revenue financed many of these adventures, remained controversial. In 87 B.C., Emperor Wudi, considered the greatest emperor of the four-century Han dynasty, died and was replaced by the eight-year-old Zhaodi. In 81 B.C., six years later, the now-teenage emperor decided, in the manner of the emperors, to invite a debate among wise men on the salt and iron monopolies. He convened sixty notables of varying points of view from around China to debate state administrative policies in front of him.

The central subject was to be the state monopolies on iron and salt. But what emerged was a contest between Confucianism and legalism over the responsibilities of good government—an expansive debate on the duties of government, state profit versus private initiative, the logic and limits of military spending, the rights and limits of government to interfere in the economy.

Though the identities of most of the sixty participants are not known, their arguments have been preserved from the Confucian point of view in written form, the *Yan tie lun*, Discourse on Salt and Iron.

On the one side were Confucians, inspired by Mencius, who, when asked how a state should raise profits, replied, "Why must Your Majesty use the word *profit*? All I am concerned with are the good and the right. If Your Majesty says, 'How can I profit my state?' your officials will say, 'How can I profit my family?' and officers and common people will say, 'How can I profit myself?' Once superiors and inferiors are competing for profit, the state will be in danger."

On the other side were government ministers and thinkers influenced by the legalist Han Feizi, who had died in 233 B.C. Han Feizi, who had been a student of one of the most famous Confucian teachers, had not believed that it was practical to base government on morality. He believed it should be based on the exercise of power and a legal code that meted out harsh punishment to transgressors. Both rewards and punishments should be automatic and without arbitrary interpretation. He believed laws should be decreed in the interest of the state, that people should be controlled by fear of punishment. If his way was followed, "the State will get rich and the army will be strong," he claimed. "Then it will be possible to succeed in establishing hegemony over other states."

In the salt and iron debate, legalists argued: "It is difficult to see, in these conditions, how we could prevent the soldiers who defend the Great Wall from dying of cold and hunger. Suppress the state monopolies and you deliver a fatal blow to the nation."

But to this came the Confucian response, "The true conqueror does not have to make war; the great general does not need to put troops in the field nor have a clever battle plan. The sovereign who reigns by bounty does not have an enemy under heaven. Why do we need military spending?"

To which came the response, "The perverse and impudent Hun has been allowed to cross our border and carry war into the heart of the country, massacring our population and our officers, not respecting any authority. For a long time he has deserved an exemplary punishment."

It was argued that the borders had become permanent military camps that caused suffering to the people on the interior. "Even if the monopolies on salt and iron represented, at the outset, a useful measure, in the long term they can't help but be damaging."

Even the need for state revenues was debated. One participant quoted Laozi, a contemporary of Confucius and founder of Daoism, "A country is never as poor as when it seems filled with riches."

The debate was considered a draw. But Emperor Zhaodi, who ruled for fourteen years but only lived to age twenty-two, continued the monopolies, as did his successor. In 44 B.C., the next emperor, Yuandi, abolished them. Three years later, with the treasury emptied by a third successful western expedition to Sogdiana in Turkistan, he reestablished the monopolies. They continued to be abolished and reestablished regularly according to budgetary needs, usually related to military activities. Toward the end of the first century A.D., a Confucian government minister had them once more abolished, declaring, "Government sale of salt means competing with subjects for profit. These are not measures fit for wise rulers."

The state salt monopoly disappeared for 600 years. But it was resurrected. During the Tang dynasty, which lasted from 618 to

907, half the revenue of the Chinese state was derived from salt. Aristocrats showed off their salt wealth by the unusual extravagance of serving pure salt at the dinner table, something rarely done in China, and placing it in a lavish, ornate saltcellar.

Over the centuries, many popular uprisings bitterly protested the salt monopoly, including an angry mob that took over the city of Xi'an, just north of Sichuan, in 880. And the other great moral and political questions of the great debate on salt and iron—the need for profits, the rights and obligations of nobility, aid to the poor, the importance of a balanced budget, the appropriate tax burden, the risk of anarchy, and the dividing line between rule of law and tyranny—have all remained unresolved issues.

~

CHAPTER TWO

Fish, Fowl, and Pharaohs

O N   T H E   E A S T E R N end of North Africa's almost unimagin-
ably vast desert, the Nile River provides a fertile green passage only a few miles wide down both banks. Egyptian civilization
has always been crammed into this narrow strip, surrounded by a
crawling wind-swept desert, like a lapping sea, that threatens to
wash it away. Even today in modern high-rise Cairo, the sweepers
come out in the morning to chase away the sand from the de-
vouring desert.

The earliest Egyptian burial sites have been found where the
desert begins at the edge of the green strip on either side. They
date from about 3000 B.C., the same time as the earliest record of
salt making in Sichuan, but before the era of the great Egyptian
states and even before such marks of Egyptian civilization as hi-
eroglyphics. The cadavers at these early burial sites still have
flesh and skin. They are not mummies and yet are surprisingly
well preserved for 5,000-year-old corpses. The dry, salty desert

sand protected them, and this natural desert phenomenon held the rudiments of an idea about preserving flesh.

To the Egyptians, a dead body was the vessel connecting earthly life to the afterlife. Eternal life could be maintained by a sculpted image of the person or even by the repetition of the deceased's name, but the ideal circumstance was to have the body permanently preserved. At all stages of ancient Egyptian civilization a tomb had two parts: one, below ground, for housing the corpse, and a second area above for offerings. In simpler burial places, the upper part might be just the open area above ground.

The upper level makes clear the importance the ancient Egyptians attached to the preparation and eating of food. Elaborate funereal feasts were held in these spaces, and copious quantities of food were left as offerings. The feasts, and sometimes the preparation of foods, were depicted on the walls. Every important period in ancient Egyptian history produced tombs containing detailed information about food. Though the intention was to leave this for the benefit of the deceased, it has given posterity a clear view of an elaborate and inventive ancient cuisine.

The poorest may have had little to eat but unraised bread, beer, and onions. The Egyptians credited onions and garlic with great medicinal qualities, believing that onion layers resembled the concentric circles of the universe. Onions were placed in the mummified cadavers of the dead, sometimes serving in place of the eyes. Herodotus, the Greek historian born about 490 B.C. and considered the founder of the modern discipline of history, described the tomb of the pyramid of Giza, built about 2900 B.C. He wrote that an inscription on one wall asserted that during twenty years of construction, the builders supplied the workers with radishes, onions, and garlic worth 1,600 talents of silver, which in contemporary dollars would be about $2 million.

But the upper classes had a richly varied diet, perhaps the most evolved cuisine of their time. Remains of food found in a tomb from before 2000 B.C. include quail, stewed pigeon, fish, ribs of beef, kidneys, barley porridge, wheat bread, stewed figs, berries, cheese, wine, and beer. Other funereal offerings found in tombs included salted fish and a wooden container holding table salt.

The Egyptians mixed brine with vinegar and used it as a sauce known as *oxalme*, which was later used by the Romans. Like the Sichuan Chinese, the Egyptians had an appreciation for vegetables preserved in brine or salt. "There is no better food than salted vegetables" are words written on an ancient papyrus. Also, they made a condiment from preserved fish or fish entrails in brine, perhaps similar to the Chinese forerunner of soy sauce.

The ancient Egyptians may have been the first to cure meat and fish with salt. The earliest Chinese record of preserving fish in salt dates from around 2000 B.C. Salted fish and birds have been found in Egyptian tombs from considerably earlier. Curing flesh in salt absorbs the moisture in which bacteria grows. Furthermore, the salt itself kills bacteria. Some of the impurities found in ancient sodium chloride were other salts such as saltpeter, which are even more aggressive bacteria slayers. Proteins unwind when exposed to heat, and they do the same when exposed to salt. So salting has an effect resembling cooking.

Whether the Egyptians discovered this process first or not, they were certainly the first civilization to preserve food on a large scale. Those narrow fertile strips on either bank of the Nile were their principal source of food, and a dry year in which the Nile failed to flood could be disastrous. To be prepared, Egyptians put up food in every way they could, including stockpiling grain in huge silos. This fixation on preserving a food supply led to considerable knowledge of curing and fermentation.

Were it not for their aversion to pigs, the Egyptians would probably have invented ham, for they salt-cured meat and knew how to domesticate the pig. But Egyptian religious leadership pronounced pigs carriers of leprosy, made pig farmers social outcasts, and never depicted the animal on the walls of tombs. They tried to domesticate for meat the hyenas that scavenged the edge of villages looking for scraps and dead animals to eat, but most Egyptians were revolted by the idea of eating such an animal. Other failed Egyptian attempts at animal husbandry include antelope, gazelle, oryx, and ibex. In the northern Sinai and what is now the southern Israeli Negev Desert, the remnants of pens for such fauna, the remains of these failed experiments, have been found. But the Egyptians did succeed in domesticating fowl— ducks, geese, quail, pigeon, and pelican. Ancient walls show fowl being splayed, salted, and put into large earthen jars.

A great source of Egyptian food was the wetlands of the Nile, the reedy marshes where fowl could be found, as well as fish such as carp, eel, mullet, perch, and tigerfish. The Egyptians salted much of this fish. They also dried, salted, and pressed the eggs of mullet, creating another of the great Mediterranean foods known in Italian as *bottarga*.

The Egyptians lay claim to another pivotal food invention: making the fruit of the olive tree edible. Almost every Mediterranean culture claims olives as its discovery. The Egyptians of 4000 B.C. believed that the goddess Isis, wife of Osiris, taught them how to grow olives. The Greeks have a similar legend. But the Hebrew word for olive, *zait*, is probably older than the Greek word, *elaia*, and is thought to refer to Said in the Nile Delta. It may have been Syrians or Cretans who first bred the *Olea europaea* from the pathetic, scraggly, wild oleaster tree. The Egyptians were not great olive oil producers and imported most of their olive oil from the Middle East. The fresh-picked fruit of the olive tree is so hard and bitter, so unappealing, that it is a wonder anyone experimented long enough to find a way to make it edible. But the Egyptians learned very early that the bitter glucides unique to this fruit, now known as oleuropeina, could be removed from the fruit by soaking in water, and the fruit could be softened in brine. The salt would render it not only edible but enjoyable.

Making olives and making olive oil are at cross purposes, since a good eating olive is low in oil content. It may be that this was characteristic of Egyptian olives. These eating olives were included in the food caches of ancient Egyptian tombs.

The Egyptians were the inventors of raised bread. To make leavened bread, a gluten-producing grain, not barley or millet, was necessary, and about 3000 B.C. the Egyptians developed wheat that could be ground and stretched into a dough capable of entrapping carbon dioxide from yeast. The starting yeast was often leftover fermented dough, sour dough, which is another example of lactic acid fermentation. Egyptian bakers created an enormous variety of breads in different shapes, sometimes with the addition of honey or milk or eggs. Most of these doughs, as with modern breads, were made from a base of flour, water, and a pinch of salt.

In 1250 B.C., when Moses liberated the Hebrew slaves, leading them out of Egypt across the Sinai, the Hebrews took with them only flat unleavened bread, *matzo,* which is described by the Hebrew phrase *lechem oni,* meaning "bread of the poor." Poor Egyptians did not have the sumptuous assortment of Egyptian raised bread but, like people outside of Egypt, ate flat bread known as *ta,* which sometimes had coarse grain, even chaff, in it and lacked the luxury of "a pinch of salt." According to Jewish legend, the fleeing

Hebrews took unleavened bread because they lacked time to let the bread rise. But it may also have been what they were used to making, or perhaps it was a conscious rejection of Egyptian culture and the luxuries of the slave owners. Raised bread and salt curing were emblematic of the high-living Egyptians.

~

THE EGYPTIANS MADE salt by evaporating seawater in the Nile Delta. They also may have procured some salt from Mediterranean trade. They clearly obtained salt from African trade, especially from Libya and Ethiopia. But they also had their own desert of dried salt lakes and salt deposits. It is known that they had a number of varieties of salt, including a table salt called "Northern salt" and another called "red salt," which may have come from a lake near Memphis.

Long before seventeenth- and eighteenth-century chemists began identifying and naming the elements of different salts, ancient alchemists, healers, and cooks were aware that different salts existed, with different tastes and chemical properties that made them suitable for different tasks. The Chinese had invented gunpowder by isolating saltpeter, potassium nitrate. The Egyptians found a salt that, though they could not have expressed it in these terms, is a mixture of sodium bicarbonate and sodium carbonate with a small amount of sodium chloride. They found this salt in nature in a *wadi*, an Arab word for a dry riverbed, some forty miles northwest of Cairo. The spot was called Natrun, and they named the salt *netjry*, or natron, after the wadi. Natron is found in "white" and "red," though white natron is usually gray and red natron is pink. The ancient Egyptians referred to natron as "the divine salt."

The culminating ritual of the lengthy Egyptian funeral was

known as "the opening of the mouth," in which a symbolic cutting of the umbilical cord freed the corpse to eat in the afterlife, just as cutting a newborn baby's cord is the prelude to its taking earthly nourishment. In 1352 B.C., the child pharoah Tutankhamen died at the age of eighteen, and his tomb, discovered in 1922, is the most elaborate and well preserved ever found. The tomb was furnished with a bronze knife for the symbolic cutting of the cord, surrounded by four shrines, each containing cups filled with the two vital ingredients for preserving mummies: resin and natron.

Investigators argue about whether sodium chloride was used in mummification. It is difficult to know, since natron contains a small amount of sodium chloride that leaves traces of common salt in all mummies. Sodium chloride appears to have been used instead of natron in some burials of less affluent people.

Herodotus, though writing more than two millennia after the practice began, offered a description in gruesome detail of ancient Egyptian mummification, which, with a few exceptions, such as his confusion of juniper oil for cedar oil, has stood up to the examination and chemical analysis of modern archaeology. The techniques bear remarkable similarity to the Egyptian practice of preserving birds and fish through disembowelment and salting:

The most perfect process is as follows: As much as possible of the brain is removed via the nostrils with an iron hook, and what cannot be reached with the hook is washed out with drugs; next, the flank is opened with a flint knife and the whole contents of the abdomen removed; the cavity is then thoroughly cleaned and washed out, firstly with palm wine and again with an infusion of ground spices. After that, it is filled with pure myrrh, cassia and every other aromatic substance, excepting frankincense, and sewn up again, after which the body is placed in natron, covered entirely over, for seventy days—never longer. When this period is over, the body is washed and then wrapped from head to foot in linen cut into strips and smeared on the underside with gum, which is commonly used by the Egyptians instead of glue. In this condition the body is given back to the family, who have a wooden case made, shaped like a human figure, into which it is put.

He then gave a less expensive method and finally the discount technique:

The third method, used for embalming the bodies of the poor, is simply to wash out the intestines, and keep the body for seventy days in natron.

The parallels between preserving food and preserving mummies were apparently not lost on posterity. In the nineteenth century, when mummies from Saqqara and Thebes were taken from tombs and brought to Cairo, they were taxed as salted fish before being permitted entry to the city.

~

MORE THAN A gastronomic development, the salting of fowl and especially of fish was an important step in the development of economies. In the ancient world, the Egyptians were leading exporters of raw foods such as wheat and lentils. Although salt was a valuable commodity for trade, it was bulky. By making a product with the salt, a value was added per pound, and unlike fresh food, salt fish, well handled, would not spoil. The Egyptians did not export great quantities of salt, but exported considerable amounts of salted food, especially fish, to the Middle East. Trade in salted food would shape economies for the next four millennia.

About 2800 B.C., the Egyptians began trading salt fish for Phoenician cedar, glass, and purple dye made from seashells by a secret Phoenician formula. The Phoenicians had built a trade empire with these products, but, in time, they also traded the products of their partners, such as Egyptian salt fish and North African salt, throughout the Mediterranean.

Originally inhabiting a narrow strip of land on the Lebanese coast north of Mount Carmel, the Phoenicians were a mixture of races, only partly Semitic. They never fused into a homogenous

Splitting and salt curing fish is illustrated in an Egyptian wall painting in the tomb of Puy-em-rê, Second Priest of Amun, circa 1450 B.C. The Metropolitan Musem of Art.

nation. Culturally, other people, first the Egyptians and later the Greeks, dominated their way of life. But economically, they were a leading power operating from major ports such as Tyre.

They traded with everyone they encountered. When Solomon constructed a temple in Jerusalem, the Phoenicians provided both wood—their famous "cedars of Lebanon"—and craftsmen. In the Old Testament it is mentioned that Jerusalem fish markets were supplied from Tyre, and the fish they sold was probably salted fish, since fresh fish would have spoiled before reaching Jerusalem.

It is a Mediterranean habit to credit great food ideas to the Phoenicians. They are said to have spread the olive tree throughout the Mediterranean. The Spanish say the Phoenicians introduced chickpeas, a western Asian bean, to the western Mediterranean, though evidence of wild native chickpeas has been found in the Catalan part of southern France. Some French writers have said the Phoenicians invented bouillabaisse, which is probably not true, and the Sicilians say the Phoenicians were the first to catch bluefin tuna off their western coast, which probably is true. The Phoenicians also established a saltworks on the western side of the island of Sicily, near present-day Trapani, to cure the catch.

Ancient Phoenician coins with images of the tuna have been found near a number of Mediterranean ports. At the time, bluefin tuna, the swift, steel-blue-backed fish that is the largest member of the tuna family, might have attained sizes of over 1,500 pounds each, but this is according to ancient writers who also believed the fish fed on acorns. Seeking warmer water for spawning, bluefin leave the Atlantic Ocean, enter the Strait of Gibraltar, pass by North Africa and western Sicily, cruise past Greece, swim through the Bosporus and into the Black Sea. At all the points of land near the bluefin's passage in the Mediterranean, the Phoenicians established tuna fisheries.

About 800 B.C., when the Phoenicians first settled on the coast of what is today Tunisia, they founded a seaport, Sfax, which still prospers today. Sfax became, and has remained, a source of salt and salted fish for Mediterranean trade. The Phoenicians also founded Cadiz in southern Spain, from where they exported tin. Almost 2,500 years before the Portuguese mariners explored West Africa, the Phoenicians sailed from Cadiz through the Strait of Gibraltar and on to the West African coast.

The Phoenicians are also credited with the first alphabet. Chinese and Egyptian languages used pictographs, drawings depicting objects or concepts. Babylonian, which became the international language in the Middle East, also had a long list of characters, each standing for a word or combination of sounds. But the Phoenicians used a Semitic forerunner of ancient Hebrew, the earliest traces of which were found in the Sinai from 1400 B.C., which had only twenty-two characters, each representing a particular sound. It was the simplicity of this alphabet as much as their commercial prowess that opened up trade in the ancient Mediterranean.

INLAND FROM THE port of Sfax are dried desert lake beds where salt can be scraped up in the dry season. This technique, the same as was used 8,000 years ago on Lake Yuncheng in China, and referred to as "dragging and gathering," was the original Egyptian way of salt gathering, the method used for harvesting natron in the wadi of Natrun. The Arabs called such a saltworks a *sebkha,* and on a modern map of North Africa, from the Egyptian-Libyan border to the Algerian-Moroccan line, from Sabkaht Shunayn to Sebkha de Tindouf, sebkhas are still clearly labeled.

In ancient times, the Fezzan region, today in southern Libya, had contact with Egypt and the Mediterranean. Herodotus wrote of the use of horses and chariots for warfare in Fezzan, which was unusual at the time. Even more unusual, horses also may have been used to transport salt. By the third century B.C., Fezzan was noted for its salt production. Fezzan producers had moved beyond simply scraping the sebkhas. The crust was boiled until fairly pure crystals had been separated, and they were then molded into three-foot-high white tapered cylinders. Traders then carried these oddly phallic objects, carefully wrapped in straw mats, by caravan across the desert. Salt is still made and transported the same way today in parts of the Sahara.

Because a profitable salt shipment is bulky and heavy, accessible transportation has always been the essential ingredient in salt trade. In most of Asia, Europe, and the Americas, waterways have been the solution. Salt was traded either through seagoing ports or, as in Sichuan, by a sprawling river system. But in the African continent, where a wealth of salt was located in the wadis and dry lake beds of the waterless Sahara, another solution was found—the camel.

The earliest known journeys across the Sahara, in about 1000 B.C., were by oxen and then by horse-drawn chariots. Trans-Saharan commerce existed in ancient times, but crossings were rare events until the third century A.D., when the camel replaced the horse. The camel was a native of North America, though it became extinct there two million years ago. Around 3000 B.C., relatively late in the history of animal domestication, camels were domesticated in the Middle East. The wild species has vanished. Between the domestication of the camel and its use in the Sahara, several millennia passed. But once the domestic camel made its Sahara debut, its use spread quickly. By the Middle Ages, caravans of 40,000 camels carried salt from Taoudenni to Timbuktu, a 435-mile journey taking as long as one month. Since then, continuing to this day, caravans of camels have moved bulk goods across the Sahara to western and central Africa. As the trade prospered, so did banditry, and the caravans grew in size for protection. As salt moved south, gold, kola nut, leather, and cotton from Hausaland, in present-day Nigeria, was traded north. Later, products for Europe, including acacia gum, which was needed for fabric sizing, and melegueta pepper, the seeds of an orange West African fruit that were a Renaissance European food craze, were also brought north. Slaves, too, were taken on this route and even at times traded for salt.

In 1352, Ibn Batuta, the greatest Arab-language traveler of the Middle Ages, who had journeyed overland across Africa, Europe, and Asia, reported visiting the city of Taghaza, which, he said, was entirely built of salt, including an elaborate mosque. By the time Europeans first discovered it in the nineteenth century, the fabled western Saharan city of salt had been abandoned. Taghaza was not the earliest report of buildings made of salt. The first-century-A.D. Roman Pliny the Elder, writing of rock salt mining in Egypt, mentioned houses built of salt.

Taghaza is imagined as a sparkling white city, but it was swept by Saharan sands, and the pockmarked salt turned a dingy gray. Though its salt construction impressed later travelers, salt blocks were the only material available for building, and Taghaza was probably a miserable work camp, inhabited mostly by the slaves forced to work it, who completely depended on the arrival of caravans to bring them food.

In ancient Taghaza, salt was quarried from the near surface in 200-pound blocks loaded on camels, one block on each side. The powerful animals carried them 500 miles to Timbuktu, a trading center because of its location on the northernmost crook of the Niger River, which connects most of West Africa. In Timbuktu, the goods of North Africa, the Sahara, and West Africa were exchanged, and the wealth from trade built a cultural center. Timbuktu became a university town, a center of learning. But to the locals in Taghaza, salt was worth nothing except as a building material. They lacked everything but salt.

It was said that in the markets to the south of Taghaza salt was exchanged for its weight in gold, which was an exaggeration. The misconception comes from the West African style of silent barter noted by Herodotus and subsequently by many other Europeans. In the gold-producing regions of West Africa, a pile of gold would be set out, and a salt merchant would counter with a pile of salt, each side altering their piles until an agreement was reached. No words were exchanged during this process, which might take days. The salt merchants often arrived at night to adjust their piles and leave unseen. They were extremely secretive, not wanting to reveal the location of their deposits. From this it was reported in Europe that salt was exchanged in Africa for its weight in gold. But it is probable that the final agreed-upon two piles were never of equal weight.

The fact that in ancient Egypt the poor were mummified with sodium chloride and the rich with natron suggests that the Egyptians valued natron more. But the reverse appears to have been true in other parts of ancient Africa. Generally, the richer Africans used salt with higher sodium chloride content, and natron was the salt of the poor. In West Africa white natron was used for bean cakes of millet or sorghum, called *kunu*. The natron in this dish was thought to be beneficial to nursing mothers. Natron was preferred to salt for bean dishes because it was thought that the carbonate counteracted gas. It was also used, and still is, as a stomach medicine—a natural bicarbonate of soda. Natron was believed to be a male aphrodisiac as well.

In Timbuktu, which was a center of not only the salt trade but the tobacco trade, a mixture of tobacco and natron was chewed. The Hausa also used natron to dissolve indigo so that the color could be fixed. Soap was made from natron and an oil from the kernel of the shea butter tree.

The African salt market has always distinguished between a wide assortment of salts, most of them impure. Salt that was mainly sodium chloride was used exclusively for eating. Sodium chloride, natron, and other salts of varying impurities, from different locations, were widely known by their own names. African merchants, healers, and cooks were well versed in this array of salts. *Trona* was the name of a well-known natron valued for food; it was gathered from the shore of Lake Chad.

Africans have maintained a tradition of a wide variety of different salts for different dishes, but they always treat any salt as a valuable substance that must not be wasted. R. Omosunlola Williams, a Nigerian educator, published a cookbook for Nigerian housewives in 1957, shortly before Nigerian independence. Among her suggestions for salt:

Salt is molded in some parts of Nigeria to make it last longer. This has to be scraped and crushed before it is used. The Yorubas use a kind of solid salt called *iyo obu*. They tie it in a piece of cloth and squeeze it in water. This is removed when it has seasoned the water sufficiently and is kept and re-used.

Africans became so accustomed to their impure salts, with specific tasks found for each blend, that when Europeans in the age of colonialism introduced pure sodium chloride, Africans mixed it with other salts to make salt compounds more to their liking.

CHAPTER THREE

Saltmen Hard as Codfish

I N 1 6 6 6, T H E *Saltzburg Chronicle* described the following incident:

In the year 1573, on the 13th of the winter month, a shocking comet-star appeared in the sky, and on the 26th of this month a man, 9 hand spans in length, with flesh, legs, hair, beard and clothing in a state of non-decay, although somewhat flattened, the skin a smoky brown color, yellow and hard like codfish, was dug out of the Tuermberg mountain 6300 shoe lengths deep and was laid out in front of the church for all to see. After a while, however, the body began to rot and was laid to rest.

He was found by salt miners in the Dürnberg mountain mine near the Austrian town of Hallein, a name which means "saltwork," near Salzburg, which means "salt town." The perfectly preserved body, dried and salted "like codfish," was that of a bearded man with a pickax found near him, evidently a miner, wearing pants, a woolen jacket, leather shoes, and a cone-shaped felt hat. The bright colors of the patterned clothing—plaid twill with brilliant red—were striking, not only because of how well the salt conserved the colors but also because Europeans are not thought of as people dressed in such a flaming palate. In 1616, a similar body had been found in nearby Hallstatt, which also means "salt town."

Inside these alpine mountains of salt, the weight of the rock overhead causes walls to shift, opening cavities and closing up shafts. Water running over the rock salt turns to brine, which then crystalizes, sealing over cracks. Three prehistoric miners have been found, trapped in their dark ancient work sites, and many tools, leather shoes, clothes in their original bright colors— the oldest color-preserved European textiles ever found—leather sacks for hauling rock salt on their backs, torches made of pine sticks bundled together and dipped in resin, and a horn possibly used to warn of cave-ins—all well preserved in salt. The bodies were dated to 400 B.C., but some of the objects found in the re- mains of a log cabin thatched-roof village on the mountainside may date back to 1300 B.C.

The colorfully dressed salt miners of Hallein were Celts. Celts did not illustrate their culture on temple walls as the Egyptians did; nor did they have chroniclers as the Greeks and Romans did. The guardians of Celtic culture, the Druids, did not leave written records. So most of what we know of them is from Greek and Ro- man historians who described the Celts as huge and terrifying men in bright fabrics. Aristotle described them as barbarians who went naked in the cold northern weather, abhorred obesity, and were hospitable to strangers. Diodorus, a Greek historian who lived in Sicily, wrote: "They are very tall in stature, with rippling muscles under clear white skin. Their hair is blond, but not naturally so. They bleach it, to this day artificially washing it in lime and combing it back from their foreheads. They look like wood demons, their hair thick and shaggy like a horse's mane. Some of these are clean shaven, but others—especially those of high rank, shave their cheeks but leave a moustache that covers the whole of the mouth and, when they eat and drink, acts like a sieve, trapping particles of food."

It is a sad fate for a people to be defined for posterity by their enemies. Even the name, Celt, is not from their own Indo-European language but from Greek. *Keltoi,* the name given to them by Greek historians, among them Herodotus, means "one who lives in hiding or under cover." The Romans, finding them less mysterious, called them Galli or Gauls, also coming from a Greek word, used by Egyptians as well, *hal,* meaning "salt." They were the salt people. The name of the town that sits on an East German salt bed, Halle, like the Austrian towns of Hallein, Swäbisch Hall, and Hallstatt, has the same root as do both Galicia in northern Spain and Galicia in southern Poland, where the town of Halych is found. All these places were named for Celtic saltworks.

Their land was in what is now Hungary, Austria, and Bavaria. The Rivers Rhine, Main, Neckar, Ruhr, and Isar are all thought to have been named by the Celts. Like the ancient Chinese emperors, they based their economy on salt and iron and so needed waterways to transport their heavy goods.

The Celts used rivers for trade and conquest. They moved west into France, south into northern Spain, and north into Belgium, named after a Celtic tribe, the Belgae. At the time that the mine shaft trapped the miner in Dürnberg, Celts were moving into the British Isles and the Mediterranean. In 390 B.C., the Celts sacked Rome, having traveled eighty miles in four days on horseback in an age when western Europeans had not seen mounted cavalry. They terrorized townspeople with their heavy swords and loud war cries. The Celts controlled Rome for the next forty years, and in 279 B.C., they invaded what is now Turkey.

Exactly how far in the world they traveled, settled, and traded is not certain. Until the nineteenth century, Western history generally dismissed the Celts as crude and frightening barbarians. But in 1846, a mining engineer named Johann Georg Ramsauer began looking for pyrite deposits in the area of the Hallstatt salt mine near Hallein. Instead he found two skeletons, an ax, and a piece of bronze jewelry. Then he discovered seven more bodies buried with valuables. He reported his findings to the government in Vienna and received funding from the curator of the imperial coin collection to continue digging. In one summer he found another 58 graves. In sixteen years he found 1,000 graves, both burials and cremation urns, and carefully cataloged thousands of objects. Numbering each grave, an artist made a watercolor record of the bodies and artifacts at each site. Ramsauer's meticulous scientific methodology made him a pioneer in the new science of archaeology. In the process, a great deal was learned about the early salt-trading Celts. The Hallstatt Period became the archaeological name for a rich early Iron Age culture, beginning about 700 B.C. and lasting until 450 B.C.

Ramsauer's Hallstatt graves were mostly from 700 to 600 B.C., with some as late as 500 B.C. The Dürnberg discoveries from 400 B.C. sugggest that the Hallstatt mine began to diminish in importance as the Dürnberg one became a more important source of salt.

Ramsauer's dig and the Dürnberg finds showed a society living off of salt mining, secluded on remote and rugged mountains at an altitude of 3,000 feet, and yet trading to the far ends of the continent. These people were buried with valuable possessions from the Mediterranean, from North Africa, even from the Near East. Ramsauer's investigation of these salt miners began to challenge the perception of northern Europe's Iron Age barbarians.

~

ONLY IN THE 1990s did Westerners become aware of the mummies that had been found in the Uyghur Autonomous Region of China. They had been discovered in and near the Tarim Basin, west of Tibet, east of Samarkand and Tashkent, between China and central Asia along the Silk Road, the principal trade route between the Mediterranean and Beijing. It was the road of Marco Polo, but these people had lived more than three millennia earlier, about 2000 B.C. As with the early Egyptian burials that are 1,000 years older, the corpses had been preserved by the naturally salty soil.

The condition of the bodies and their bright colored clothing was spectacular. The men wore leggings striped in blue, ochre, and crimson. They appeared to be tall with blond or light brown hair, sometimes red beards, and the women's hair woven in long blond braids. These unknown people were in appearance notably similar to the large blue-eyed blond Celtic warriors described by the Romans almost two millennia later. Their conical felt hats and twill jackets bore a close resemblance to those of the salt miners in Hallein and Hallstatt—not unlike the much later plaids of the Scottish Highlands. The red-and-blue pinstripes were almost identical to fabrics found in the Dürnberg mine. Textile historian Elizabeth Wayland Barber concluded that even the weave was nearly identical workmanship. Why Celts might have been in the salty desert of Asia many centuries before there were known to be Celts remains a mystery.

In the centuries when the Celtic culture was documented, beginning 1,300 years after these seemingly Celtic bodies were buried in Asian salt, they did trade and travel great distances, usually selling salt from their rich central European mines. Like the Egyptians, they learned that it was not as profitable to trade and transport salt as salted foods.

According to the Greeks and Romans, who not only wrote about Celts but traded Mediterranean products for their salt and salt products, Celts ate a great deal of meat, both wild and domesticated. Salted meat was a Celtic specialty.

When the Romans finally succeeded in imposing their culture on the Celts, Moccus, which means "pig," was the Celtic name for the god Mercury. The Celts did not mean it unkindly. To the pig-loving Celts, the leg of wild boar was considered the choicest piece of meat and was reserved for warriors. With domesticated pigs also, according to Strabo, the first-century-B.C. Greek historian, the Celts preferred the legs. It is likely that among the Celtic contributions to Western culture are the first salt-cured hams.

Athenaeus, a first-century-A.D. Greek living in Rome, wrote that the Celts most valued the upper part of the ham, which was reserved for the bravest warrior. If two warriors claimed rights to this cut, the dispute would be settled by combat. Fighting over the ham may be more the Greco-Roman view of Celts than the reality. But the Celts certainly made, traded, and ate hams.

Among the few remaining Celtic cultures, this tradition of savoring a salt-cured leg from the hunt endures. An example is Scottish salted haunch of venison.

Take the venison to be salted after it has hung in the larder for two days. Cut it into pieces the required size. See that it is clean and free from fly, *but on no account wash it with water* [her italics]. Take 2 pounds kitchen salt, one quarter pound demerara sugar, 1 teaspoon black pepper, one half teaspoon nitre [natron].

Mix these well together. Rub pieces of venison on every side with this mixture for 2–3 days in succession. Then place them in a wooden tub, or earthenware jar, and press them well together. After 10 days the venison is ready for use. Venison treated in this way, if pressed into a jar and the air excluded, should keep for months, and a haunch which has been well salted in this manner for about three weeks can be hung up to dry as a ham.—*Margaret Fraser,* A Highland Cookery Book, *1930*

According to Annette Hope, an Edinburgh librarian who collected Scottish recipes, Margaret Fraser came from a family of gamekeepers on a Highland estate, and most of her recipes were for venison, though the same ideas may have been used for legs of other game and domestic meat. The sugar—she specified the light brown of Demerara, British Guiana—would not have been used by original Celts, but the natron may have been.

～

THE EARLY CELTIC salt miners understood their mountains. They realized that horizontal shafts from the mountainside, though a great deal easier to travel and move rock through, would require far more digging to reach the rich salt deposits. Instead they dug at steep angles and skillfully shored up the shafts. The miners had to climb out, flaming torch clenched in their teeth, leather back sack loaded with rock, at forty-five- or fifty-degree angles. Though the master ironworkers of their age, they made their picks and other metal tools out of bronze, the antiquated metal of a more primitive era. They seem to have learned that bronze would not be corroded by salt the way iron is.

The Celts, or their central European ancestors known as the Urnfield people, because they cremated their dead and buried them in urns, had many innovations besides those in salt mining. They developed the first organized agriculture in northern Europe, experimenting with such revolutionary ideas as fertilizer and crop rotation. They introduced wheat to northern Spain. They were sophisticated bronze casters, skilled iron miners and forgers. They introduced to much of western Europe iron and their many iron inventions, including chain armor and the feared Celtic sword, which was three feet long. But they also invented the seamless iron rim for wagon wheels, the barrel, and possibly the horseshoe. They may have been the first Europeans to ride horses.

One thing the Celts were not advanced in was statecraft. Ironically, the closest the Celts ever came to fusing into a nation was in the first century when Julius Caesar conquered Gaul. A Celtic leader named Vercingetorix, which means "warrior king," gathered warriors from the diverse Celtic groups to face the Romans at Alesia, now Alise-Sainte Reine on the lower Seine. Vercingetorix's father had attempted the same thing unsuccessfully in 80 B.C.

According to Caesar, while besieged Celts were so desperate that they were debating whether to eat the elderly noncombatants, forty-one Celtic tribes responded to Vercingetorix's call by sending a relief column to Alesia of 8,000 horsemen and 250,000 foot soldiers.

Some historians believe that had the Celts won at Alesia, it would have been the beginning of a united Celtic nation. But the Romans won and subjugated the Celts and wrote their history.

Despite the fame of their bright clothing, Celts are described going naked into battle except for horned helmets. We are told that they had frightening war cries and that the terrifying songs of their ancestors were preludes to violent attacks. They fought, the Romans said, with a furor. And they swooped off heads with their large iron swords and hung these trophies on their houses

Gold stater of the Avernes tribe with the face of Apollo and the inscription "Vercingetorixs." The Granger Collection.

or strung them along the horse bridle. Vercingetorix was apparently a ruthless leader, a fanatic obsessed with freeing his people from the Romans, willing to destroy entire towns and ruthlessly level opponents to achieve his goal.

But he was trying to stand up to the Roman legions of Julius Caesar. The Roman historian Plutarch estimated that the civilized Romans under Julius Caesar, in his decade-long campaign in Gaul, destroyed 800 towns and villages and enslaved 3 million people.

After the Roman campaigns were over, all that remained of Celtic life were isolated groups on the far Atlantic coasts: northwestern Iberia, the Brittany peninsula, the Cornish tip of England, Wales, Ireland, Scotland, and the Isle of Man. All of these groups were treated by the chroniclers of later nation-states as recalcitrant people interfering with the building of great states—Britain, France, or Spain.

The Roman victory had been total. Celtic inventions—in salt mining, iron, agriculture, trade, horsemanship—enriched the Roman Empire. Celtic salt mines became part of Roman wealth, and Celtic hams became part of the Roman diet with few ever remembering that such things were once Celtic. The Celts were innovators. The Romans were nation builders.

CHAPTER FOUR

∿

Salt's Salad Days

THE ROMANS PAID homage to democracy, the rights of the common citizen and, for a time, republicanism. But they rarely lived up to any of these ideals. Roman history is the chronic struggle between the privileged patricians and the disenfranchised plebeians. Plebeians fought to have a voice, and patricians endeavored to keep them excluded. The Roman patrician often tried to keep his privileges by offering lesser rights to plebeians. In this spirit, patricians insisted that every man had a right to salt. "Common salt," as it has come to be known, was a Roman concept.

Patricians ate an elaborate cuisine that expressed opulence in ingredients and presentation. Roman cooks seemed to avoid leaving anything in its natural state. They loved the esoteric, such as sow's vulva and teats, a dish that is frequently mentioned for banquets and which provoked a debate as to whether it should be from a virgin sow or, as Pliny the Elder suggested, one whose first litter was aborted.

Sometimes the cuisine emphasized local pride. The best pike had to be caught in the Tiber between the city bridges of Rome. But food was also a way to boast of conquest, with hams from Germania, oysters from Britannia, and sturgeon from the Black Sea. Meanwhile, plebeians ate coarse bread, cereals, a little salt fish, and olives. And the government made certain they had salt.

Roman government did not maintain a monopoly on salt sales as did the Chinese, but it did not hesitate to control salt prices when it seemed necessary. The earliest record of Roman government interference in salt prices was in 506 B.C., only three years before the kingdom was declared a republic. The state took over Rome's premier source of salt, the private saltworks in Ostia, because the king regarded its prices as too high.

Both under the republic and, later, the empire, Roman government periodically subsidized the price of salt to ensure that it was easily affordable for plebeians. It was a gift, like a tax cut, that government could bestow when in need of popular support. On the eve of Emperor Augustus's decisive naval campaign defeating Mark Anthony and Cleopatra, Augustus garnered public support by distributing free olive oil and salt.

But during the Punic Wars (264 to 146 B.C.), a century-long struggle-to-the-death for control of the Mediterranean with the Phoenician colony of Carthage, Rome manipulated salt prices to raise money for the war. In the fashion of the Chinese emperors, the Roman government declared an artificially high price for salt and put the profits at the disposal of the military. A low price was still maintained in the city of Rome, but elsewhere a charge was added in accordance with the distance from the nearest saltwork. This salt tax system was devised by Marcus Livius, a tribune, a government official representing plebeians. Because of his salt price scheme, he became known as the *salinator*, which later became the title of the official in the treasury who was responsible for decisions about salt prices.

⌒

MOST ITALIAN CITIES were founded proximate to saltworks, starting with Rome in the hills behind the saltworks at the mouth of the Tiber. Those saltworks, along the northern bank, were controlled by Etruscans. In 640 B.C., the Romans, not wanting to be dependent on Etruscan salt, founded their own saltworks across the river in Ostia. They built a single, shallow pond to hold seawater until the sun evaporated it into salt crystals.

The first of the great Roman roads, the Via Salaria, Salt Road, was built to bring this salt not only to Rome but across the interior of the peninsula. This worked well in the Roman part of the Italian peninsula. But as Rome expanded, transporting salt longer distances by road became too costly. Not only did Rome want salt to be affordable for the people, but, more importantly as the Romans became ambitious empire builders, they needed it to be available for the army. The Roman army required salt for its soldiers and for its horses and livestock. At times soldiers were even paid in salt, which was the origin of the word *salary* and the expression "worth his salt" or "earning his salt." In fact, the Latin word *sal* became the French word *solde*, meaning pay, which is the origin of the word, soldier.

To the Romans, salt was a necessary part of empire building. They developed saltworks throughout their expanded world, establishing them on seashores, marshes, and brine springs throughout the Italian peninsula. By conquest they took over not only Hallstatt, Hallein, and the many Celtic works of Gaul and Britain but also the saltworks of the Phoenicians and Carthaginians in North Africa, Sicily, Spain, and Portugal. They acquired Greek works and

Black Sea works and ancient Middle Eastern works including the saltworks of Mount Sodom by the Dead Sea. More than sixty salt-works from the Roman Empire have been identified.

Romans boiled sea salt in pottery, which they broke after a solid salt block had formed inside. Piles of pottery shards mark many ancient Roman sea salt sites throughout the Mediterranean. The Romans also pumped seawater into single ponds for solar evaporation, as in Ostia. They mined rock salt, scraped dry lake beds like African sebkhas, boiled the brine from marshes, and burned marsh plants to extract salt from the ashes.

None of these techniques were Roman inventions. Aristotle had mentioned brine spring evaporation in the fourth century B.C. Hippocrates, the fifth-century-B.C. physician, seems to have known about solar-evaporated sea salt. He wrote,

> The sun attracts the finest and lightest part of the water and carries it high up; the saltiness remains because of its thickness and weight, and in this way the salt originates.

The Roman genius was administration—not the originality of the project but the scale of the operation.

⌒

THE ROMANS SALTED their greens, believing this to counteract the natural bitterness, which is the origin of the word *salad*, salted. The oldest surviving complete book of Latin prose, Cato's second-century-B.C. practical guide to rural life, *De agricultura*, suggests eating cabbage this way:

> If you want your cabbage chopped, washed, dried, sprinkled with salt or vinegar, there is nothing healthier.

Salt was served at the table, in a simple seashell at a plebeian's table or in an ornate silver saltcellar at a patrician's feast. In fact, since salt symbolized the binding of an agreement, the absence of a saltcellar on a banquet table would have been interpreted as an unfriendly act and reason for suspicion.

Cato suggested testing brine for sufficient salinity to use in pickling by seeing if an anchovy or an egg would float in it. The anchovy test has not endured, but the egg test remains the standard household technique throughout the Mediterranean. In northern Europe a floating potato is sometimes used.

Most of the salt consumed by Romans was already in their food when they bought it at the market. Salt was even added to wine in a spicy mixture called *defrutum,* which, in the absence of bottling corks, was used to preserve the wine. This may explain why their food was said to be extremely salty and yet the consumption of table salt was not remarkable. In the first century A.D., Pliny estimated that the average Roman citizen consumed only 25 grams of salt a day. The modern American consumes even less if the salt content of packaged food is not included.

The Romans used a great deal of salt in the hams and other pork products that they seemed to have learned about from the Celts. Sausages—pork and other meats, preserved in salt, seasoned and stuffed in natural casings from intestines, bladders, or stomachs—were both imported from Gaul and made locally. The recipes for many of the French and Italian sausages of today date from Roman times.

Originally, hams and sausages were brought to Rome from the conquered northern empire. According to Strabo, the well-traveled first-century-B.C. Greek historian, the most prized ham in Rome came from the forests of Burgundy. At the time those forests were Celtic, but the French, who have a habit of claiming

Celtic history—they have made Vercingetorix a French national hero—insist that ham is a French invention, albeit from Celtic Gaul. But the Romans were importing ham from numerous Celtic regions, including what is present-day Germany. The hams of Westphalia, which were dried, salted, and then smoked with unique local woods—a recipe still followed today in Westphalia—were very popular with Romans.

Cato, like many Romans, was a ham enthusiast. In fact, at a time when Romans often took family names from agriculture, Cato was called Marcus Porcius. Porky Marcus's recipe for mothproof ham was an attempt to produce a Westphalia-type product. The addition of oil and vinegar was intended to reproduce the savage taste of the wild north.

> After buying legs of pork cut off the feet. 1/2 peck ground Roman salt per ham. Spread the salt in the base of a vat or jar, then place a ham with the skin facing downwards. Cover completely with salt. Then place another above it and cover in the same way. Be careful not to let meat touch meat. Cover them all in the same way. When all are arranged, cover the top with salt so that no meat is seen, and level it off.
>
> After standing in salt for five days, take all hams out with the salt. Put those that were above below, and so rearrange and replace. After a total of twelve days take out the hams, clean off all the salt and hang in the fresh air for two days. On the third day clean off with a sponge, rub all over with oil, hang in smoke for two days. On the third day take down, rub all over with a mixture of oil and vinegar and hang in the meat store. Neither moths nor worms will attack it.—*Cato, De agricultura, second century* B.C.

OLIVES, PRESERVED IN salt, along with the older idea, crushed into oil, were staples of the Roman diet and a basic food of the working class. Patricians ate olives at the beginning of a meal. For plebeians, they were the meal. Cato listed his workers' provisions as bread, olives, wine, and salt. Despite their hardness, olives must be hand-picked because any bruising can be ruinous in the pickling process. Harvesting olives requires so much care that in ancient times it was believed that conditions were only auspicious for a successful harvest during the last quarter moon of each month.

The bruised or damaged fruit was cured for the workers, but the successfully gathered olives were cured in a variety of ways for sale. Apicius, the great Roman food writer, mentioned an olive called *columbades* that was cured in seawater.

The Romans preserved many vegetables in brine, sometimes with the addition of vinegar, including fennel, asparagus, and cabbage. Cato's 2,200-year-old recipe, using repeated soakings to remove the oleuropeina, and then salt for lactic acid fermentation, is still one of the standard techniques. When he said "soaked sufficiently," he neglected to mention that this takes days.

How green olives are conserved.

Before they turn black, they are to be broken and put into water. The water is to be changed frequently. When they have soaked sufficiently they are drained, put in vinegar, and oil is added. ½ pound salt to 1 peck olives. Fennel and lentisk [the seeds of the lentisk tree] are put up separately in vinegar. When you decide to mix them in, use quickly. Pack in preserving-jars. When you wish to use, take with dry hands.

FISH WAS THE centerpiece of Roman cuisine. When salted, it was also at the heart of Roman commerce. The Greek physician Galen, who lived from A.D. 130 to 200, wrote about the Roman salt fish trade. Galen was the first to understand the significance of reading pulses, and his writings on health and diet were a major influence on medicine well into the Middle Ages. It was not a coincidence that a physician would be writing about salt fish since, like salt, it was considered both a food and a medicine.

Galen described Rome's ports busy with ships unloading salt fish from the eastern and western Mediterranean. He said that the best salt fish he knew was called *sarda*, but he also praised the tuna salted in Sardinia or in Gades, Spain, and salted mullet from the Black Sea. Sarda may refer to the small tuna now called bonito or Atlantic mackerel, or the sardine, a small young pilchard, which is a uniquely European fish. He also praised salt fish from Egypt and cured Spanish mackerel from the port of Sexi in southern Spain.

By Galen's time, the centuries-old trade in both salted fish and fermented salt fish sauce had been well established in the Mediterranean. It had even been a topic of physicians before. But what struck Galen in the second century was that never before in history had the trade been so extensive and on such a massive scale.

IN 241 B.C., at the end of the Punic Wars, when Phoenician Carthage was crushed, Sicily, the largest island in the Mediterranean, came under Roman control. Sicily was known as the "breadbasket of Rome" for its grain. But it also had valuable fisheries. Catching, salt curing, and selling fish was the major activity of the entire Sicilian coastline, and the most famous fish throughout the Mediterranean was the salted bluefin tuna.

The Sicilians made salt by boiling the seawater caught in the island's many marshes. Excavations have revealed ancient salt-works concentrated in the western part of the island around Trapani and on the island of Favignana. Not coincidentally, these are the areas from which the bluefin tuna is fished.

Archestratus, the Sicilian-born fourth-century-B.C. Greek poet and gourmet, praised his native island's tuna, both fresh and salt-ed, stored in jars. Normally when a tuna was caught, the choice upper body parts were eaten fresh, and the drier tail meat was reserved for salting. But Archestratus offered an interesting compromise.

> Take the tail of the female tuna—and I'm talking of the large female tuna whose mother city is Byzantium. Then slice it and bake all of it properly, simply sprinkling it lightly with salt and brushing with oil. Eat the slices hot, dipping them into a sharp brine. They are good if you want to eat them dry, like the immortal gods in form and stature. If you serve it sprinkled with vinegar, it will be ruined.—*Archestratus,* The Life of Luxury, *fourth century B.C.*

Archestratus also admired the Black Sea tuna coming from Byzantium, the site of present-day Istanbul. These fish were from the same schools. The bluefin passes Sicily on its spawning journey to the Black Sea. In pre-Roman times, the Black Sea was a major fishing and salt fish area, especially for tuna, but also herring, sturgeon, flounder, mackerel, and anchovies. Herodotus singled out the salted sturgeon of what is now the Dnieper River, which flows through the Ukraine into the Black Sea.

⌒

FROM THE BLACK Sea to the Strait of Gibraltar, salt production was usually placed near fishing areas, creating industrial zones that produced a range of salt-based products, including various types of salt fish, fish sauces, and purple dye.

Salsamentum, from *sal,* salt, was the Roman word for salted products. The most commercially important salsamentum was salt fish. Whereas the Greeks had developed an entire vocabulary for salt fish, describing the type of cure, the place of origin, the cut of fish, salted with scales, or without scales, the Romans simply spoke of salsamentum, from which they made a good deal of money.

After the producers made all of these salsamenta, the scraps— the innards, the gills, and the tails—were used to make sauce. Roman writings mention four classes of sauce: *garum, liquamen, allec,* and *muria.* The exact meaning of these terms has been lost. Allec may have been the leftover sludge after the sauce was strained. Garum and liquamen ended up being generic terms for fermented fish sauce.

To make the sauce, the fish scraps were put in earthen jars with alternating layers of salt and weighted on the top to keep them submerged in the pickle that developed as salt drew moisture out of the fish. Classics scholars have searched for precise ancient garum recipes, but the clearest are medieval, from *Geoponica,* a Greek agricultural manual written about A.D. 900. It offered a number of garum recipes based on earlier sources:

The so-called liquamen is made in this manner: the intestines of fish are thrown into a vessel and salted. Small fish, either the best smelt, or small mullet, or sprats, or wolffish, or whatever is deemed to be small, are all salted together and, shaken frequently, are fermented in the sun.

After it has been reduced in the heat, garum is obtained from it in this way: a large, strong basket is placed into the vessel of the aforementioned fish, and the garum streams into the basket. In this way the so-called liquamen is strained through the basket when it is taken up. The remaining refuse is allec. . . .

Next, if you wish to use the garum immediately, that is to say not ferment it in the sun, but to boil it, you do it this way. When the brine has been tested, so that an egg having been thrown in floats (if it sinks, it is not sufficiently salty), and throwing the fish into the brine in a newly-made earthenware pot and adding in some oregano, you place it on a sufficient fire until it is boiled, that is until it begins to reduce a little. Some throw in boiled-down must [unfermented wine]. Next, throwing the cooled liquid into a filter, you toss it a second and a third time through the filter until it turns out clear. After having covered it, store it away.

Physicians saw in garum all of the health benefits of salt fish contained in a bottle. It was prescribed as a medicine or, more commonly, mixed with other ingredients to make a medicine, usually for digestive disorders, and for such problems as sores, for which salt has clear healing powers. But it was also prescribed for a range of other ailments, including sciatica, tuberculosis, and migraine headaches.

The only other place in the ancient world to use garum was Asia. The sauce appears to be, as some historians believe of the domesticated pig, an idea that occurred independently to the East and the West. The Asian sauce is thought to have originated in Vietnam, though the Vietnamese must have taken it in ancient times from the Chinese soy sauce, in those early times when the Chinese fermented fish with the beans.

In Vietnam salt is so appreciated that poor people sometimes make a meal of nothing more than rice and a salt blend, either salt and chili powder or the more expensive salt with ground, grilled sesame seeds. Salt is also mixed with minced ginger root. But more popular than any of these, since ancient times, is *nước mám*, a brine made from salting small fish. Unlike the Roman version, Asian garum has remained popular into modern times and is made virtually everywhere in Southeast Asia, including Cambodia, Myanmar, Laos, and the Philippines, where it is called *bagoong*. In Thailand, where it is called *nam pla,* it is produced by more than 200 factories. The Koreans, the Chinese, the Japanese, even the Indians have variations.

In Vietnam, *nước mắm* is served over fruits and vegetables with hot peppers and garlic for New Year. *Tré* is a boiled pig head cut in slices and seasoned with *nước mắm*. *Nước mắm* has many variations: *mắm cáy* uses crab; *mắm mức*, squid; *mắm tôm*, shrimp.

The French, when they first encountered this sauce, apparently forgetting their own Latin heritage, were horrified that the Vietnamese ate "rotten fish." The Romans had encountered similar reactions. In the early twentieth century, the celebrated Institut Pasteur in Paris studied *nước mắm* for sixteen years, from 1914 to 1930, to understand the fermentation process that Vietnamese peasants had been employing for centuries. The two necessary ingredients were fish and salt. The fish were usually small ones of the Clupeidae family, to which herrings and sardines belong. The fish sat in salt for three days, which produced a juice, some of which was reserved to ripen in the sun, while the remainder was pressed with the fish to produce a mush. The two were then mixed together and left for three months, sometimes much longer. Then the solid parts were strained out.

⌒

THE ROMANS USED garum in much the same way that the Chinese used soy sauce. Rather than sprinkling salt on a dish, a few drops of garum would be added to meat, fish, vegetables, or even fruit. The oldest cookbook still in existence, *De re coquinaria*—which is credited to Apicius, though it appears to be a compilation of a number of Roman cooks from the first century A.D.—gives far more recipes with garum than with salt. Garum was much more expensive than salt, but Apicius was clearly writing for the upper classes. According to Seneca, Apicius committed suicide because, having spent one tenth of a considerable fortune on his kitchen, he realized that he could not long continue in the style he had chosen.

The following recipe from Apicius is an example of the kind of elaborate molded dish that the Romans loved. It is seasoned with garum, and there is no other mention of salt in the dish.

[Place cooked] mallows, leeks, beets, or cooked cabbage sprouts, roasted thrushes and quenelles of chicken, tidbits of pork or squab, chicken, and other similar shreds of fine meats that may be available. Arrange everything in alternating layers [in a mold].

Crush pepper and lovage [a bitter herb, common as parsley in ancient Rome] with two parts old wine, one part broth [garum], one part honey and a little oil. Taste it; and when well-mixed and in due proportions put in a sauce pan and allow to heat moderately; when boiling add a pint milk in which [about eight] eggs have been dissolved; pour over [the mold and heat slowly but do not allow to boil] and when thickened serve. [The dish would usually be unmolded before serving.]

A simpler recipe using garum instead of salt is that for braised cutlets:

Place the meat in a stew pan, add one pound of broth [garum], a like quantity of oil, a trifle of honey, and thus braise.

And here is one for a fish sauce:

Sauce for roasted red mullet: pepper, lovage, rue [an aromatic evergreen], honey, pine nuts, vinegar, wine, garum, and a bit of oil. Heat and pour over the fish.

Although this style of cooking was a kind of haute cuisine for the elite, costly garum was frequently described as "putrid," which is to say rotten. "That liquid of putrefying matter," said Pliny. Seneca, the outspoken first-century philosopher, called it "expensive liquid of bad fish." But his protégé, the poet Martial, apparently did not agree since he once sent garum with the note "accept this exquisite garum, a precious gift made with the first blood spilled from a living mackerel."

But Martial was probably writing about *garum sociorum,* which means "garum among friends," the most expensive garum, made exclusively from mackerel in Spain. Garum factories of varying standards were built on the coast not only in Roman ports such as Pompeii, but in southern Spain, the Libyan port of Leptis Magna, and in Clazomenae in Asia Minor. Since the Britons both made salt and exported fish, it is likely that England too was involved in the Roman salt fish and garum trade.

Many types of garum were made—even a kosher garum, *garum castimoniale,* for the sizable Jewish market in Roman-occupied Israel. Castimoniale, in accordance with Jewish dietary law, was guaranteed to have been made only from fish with scales. The usual fish for garum—tuna, sardines, anchovies, or mackerel—all have scales and are kosher. But it seems even in the first century, a rabbinical certification brought a better price.

As the market for garum grew, low-priced brands began to appear on the market. Slaves even made garum from household fish scraps. There is often a thin line between pungent and rotten, and some of these sauces must have emitted sickening smells. Apicius offered a recipe for fixing garum that smelled bad.

> If garum has contracted a bad odor, place a vessel upside down and fumigate it with laurel and cypress and before ventilating it, pour the garum in the vessel. If this does not help matters, and if the taste is too pronounced, add honey and fresh spikenard [new shoots—*novem spicum*] to it; that will improve it. Also new must should be likewise effective.

WHEN THE ROMANS took over the Phoenician salt fish trade, they discovered how to make their purple dye. A logical by-product of fish salting, the dye was produced by salting murex, a Mediterranean mollusk whose three-inch shell resembles a dainty whelk.

According to legend, the presence of this dye was discovered when Hercules took his sheepdog for a walk along the beach in Tyre. When the inquisitive dog bit into a shellfish, his mouth turned a strange dark color. From at least as early as 1500 B.C., this dye brought wealth to merchants in Tyre.

murex

The painstakingly extracted purple dye was a luxury item of such prestige that the color purple became a way of showing wealth and power. Julius Ceasar decreed that only he and his household could wear purple-trimmed togas. The high priests of Judaism, the Cohanim, dyed the fringes of their prayer shawls purple. Cleopatra dyed the sails of her warship purple. Virgil, the first-century-B.C. poet, wrote, "And let him drink from a jeweled cup and sleep on Sarran purple," Sarran meaning "from Tyre."

Pliny wrote that men were slaves to "luxury, which is a very great and influential power inasmuch as men scour forests for ivory and citrus-wood and all the rocks of Gaetulia (North Africa) for the murex and for purple."

Romans who could afford it also ate murex, the ultimate luxury food, which they called "purple fish." One recipe called for it to be served surrounded by the tiny birds known as figpeckers. It is still eaten, steamed and twisted out of the shell with a pin, by the French, who call it *rocher*, the Spanish, who call it *cañadilla*, and the Portuguese, who call it *búzio*.

Pliny described the arduous process to obtain the dye:

There is a white vein with a very small amount of liquid in it: . . . Men try to catch the murex alive because it discharges its juice when it dies. They obtain the juice from the larger purple-fish by removing the shell: they crush the smaller ones together with their shell, which is the only way to make them yield their juice. . . .

The vein already mentioned is removed, and to this, salt has to be added in the proportion of about one pint for every 100 pounds. It should be left to dissolve for three days, since, the fresher the salt, the stronger it is. The mixture is then heated in a lead pot with about seven gallons of water to every fifty pounds and kept at a moderate temperature by a pipe connected to a furnace some distance away. This skims off the flesh which will have adhered to the veins, and after about nine days the cauldron is filtered and a washed fleece is dipped by way of a trial. Then the dyers heat the liquid until they feel confident of the result.—*Gaius Plinius Secundus, Pliny the Elder,* Historia naturalis, *first century* A.D.

The nature of the precious liquid from which purple came would not be entirely understood for another two millennia. In 1826, a twenty-three-year-old student at the Ecole de Pharmacie, Antoine Jérôme Balard, after studying the composition of salt marshes, concluded that the blackish-purplish, foul-smelling liquid present in marsh water, the residue water from which salt crystals had formed, was a previously unidentified chemical element. Because the liquid was identical to the purple secretion of the murex, he named the new element *muride*. The *Académie Française*, wary of having major discoveries come from students, thought at the least it should not let him give the name. So they changed *muride* to *bromine*, a word meaning "stench."

Murex was made in much of the Roman Mediterranean, in North Africa, on the Mediterranean coast of Gaul. Mountains of ancient murex shells from Roman times have been found in the Israeli port of Accra. Between this stinking bromine solution from the dyeworks and the smell of fish being cured, the Roman Empire must have had a redolent coast.

AFTER THE FALL of Rome in the fifth century, garum was often thought of as just one of the unpleasant hedonistic excesses for which Rome was remembered. Leaving fish organs in the sun to rot was not an idea that endured in less extravagant cultures. Of course when garum was made properly, the salt prevented rotting until the fermentation took hold. But it became increasingly difficult to convince people of this. Anthimus, living in sixth-century Gaul, in a culture that was leaving Rome behind, rejected garum for salt or even brine:

> Loin of pork is best eaten roasted, because it is a good food and well digested, provided that, while it is roasting, it is spread with feathers dipped in brine. If the loin of pork is rather tough when eaten, it is better to dip in pure salt. We ban the use of fish sauce from every culinary role.—*Anthimis,* De obseruatione ciborum (*On the Observance of Foods*), circa A.D. 500

Anthimus's pronouncement on garum has echoed through Western cooking: "*Nam liquamen ex omni parte prohibemus*," We ban the use of garum from every culinary role.

Sardines, which got their name from being a highly praised salt fish cured in Sardinia, were favored for garum. Gargilius Martialis, writing in the third century A.D., specified sardines for making garum. Modern divers examining a shipwreck off of southeastern Sicily found fifty Roman amphorae containing salted sardines. But in later centuries, sardines became better appreciated fresh with a sprinkling of salt.

> Sardines: In their natural state they should be fried; when done, garnish them with orange juice and a little of the frying

oil and salt; they are eaten hot.—Cuoco Napoletano, *anonymous, Naples, late 1400s*

After the fall of Rome, garum vanished from the Mediterranean, the region lost its importance as a salt fish producer, and the purple dye industry faded. But the Roman idea that building saltworks was part of building empires endured.

CHAPTER FIVE

⌒

Salting It Away in the Adriatic

THE FALL OF the Roman Empire left the Mediterranean, the most economically important region of the Western world, without a clear leader but with many aspirants. The region was the most competitive it had been since the rise of the Phoenicians.

The entire coast of the Mediterranean was studded with saltworks, some small local operations, others big commercial enterprises such as the ones in Constantinople and the Crimea. The ancient Mediterranean saltworks that had been started by the Phoenicians, like power itself, passed from Romans to Byzantines to Muslims.

The saltworks that the Romans had praised remained the most valued. Egyptian salt from Alexandria was highly appreciated, especially their *fleur de sel,* the light crystals skimmed off the surface of the water. Salt from Egypt, Trapani, Cyprus, and Crete all had great standing because they had been mentioned by Pliny in Roman times.

VENICE, THE ONE Italian city that was not part of Roman history, was settled on islands in the lagoons in the Adriatic. The coast of Venetia was substantially different than it is today. A series of sandbars, called *lidi*, sheltered lagoons from the storms of the Adriatic. These lagoons stretched from Ravenna, the commercial and political center of the Venetian coast, up the estuary of the Po River to Aquileia, on the opposite side of the Adriatic next to Trieste. Lido has become the name of a sandbar in Venice, particularly popular with the hordes of tourists who now wander the streets and canals of that city. But even in Roman times, the lidi were for tourists, summer resorts for affluent Romans.

In the sixth century, the mainland, what had been the Roman province of Veneto, was invaded by Germanic tribes. To preserve their independence, small groups of people, like Bostonians fleeing to Martha's Vineyard, moved to the islands protected by their summer vacation lidi.

Cassiodorus, a sixth-century high Roman official turned monastic scholar, admired these settlements in the lagoons. He likened their houses, part on land and part on sea, to aquatic birds.

> Rich and poor live together in equality. The same food and similar houses are shared by all; wherefore they cannot envy each other's hearths, and so they are free from the vices that rule the world. All your emulation centers on the saltworks; instead of ploughs and scythes, you work rollers [for salt production] whence comes all your gain. Upon your industry all other products depend for, although there may be someone who does not seek gold, there never yet lived the man who does not desire salt, which makes every food more savory.—*Cassiodorus*, A.D. 523.

As with Rome, Venetian democracy was more of an ideal than a practice. But, though Cassiodorus may have been overly enthusiastic about Venetian egalitarianism, the importance that he attributed to salt in Venice was not exaggerated. Salt was the key to a policy that made Venice the dominant commercial force of southern Europe.

The Italian mainland was originally much farther away from the islands that are now the city of Venice. The area between these islands and the peninsula of Comacchio was called the Seven Seas. "To sail the seven seas" meant simply sailing the Seven Seas—accomplishing the daunting task of navigating past the sandbars of those treacherous twenty-five miles.

About A.D. 600, Venetians started using landfill to extend the mainland closer to the islands of modern-day Venice. The Seven Seas became a landmass with a port named Chioggia. Below it, in a now much-narrowed lagoon, was Comacchio, overlooking the delta of the Po. Ravenna, formerly a port, became an inland city, and nearby Cervia became its port.

By the seventh century, with the Seven Seas gone, the Venetians built salt ponds along the newly formed land of Chioggia. Cassiodorus wrote that the Venetians were using "rollers," but sometimes this is translated as "tubes" or "cylinders." It is not clear if he was speaking of rollers to smooth down the floors of artificial evaporation ponds or cylindrical pottery to boil seawater into crystals. Both techniques had been common in Rome.

Between the sixth and ninth centuries, the last great technical advance in salt manufacturing until the twentieth century was invented. Instead of trapping seawater in a single artificial pond, closing it off, and waiting for the sun to evaporate the water, the salt makers built a series of ponds. The first, a large open tank, had a system of pumps and sluices that moved the pond's seawater to the next pond, after it reached a heightened salinity. There the water evaporated further, and a still denser brine was moved to the next pond. At the same time, fresh seawater was let into the first pond so that a fresh batch of brine was always beginning. When brine reaches a sufficient density, salt precipitates out—it crystalizes, and the crystals fall to the bottom of the pond, where they can be scooped out. In a pond with only solar heat, it may take a year or more for seawater to reach this density. But given sufficient sun and wind and a season dry enough not to have rainfall dilute the ponds, the only limit to production is the available area, the number of ponds that can operate simultaneously. It requires little equipment, a very small investment, and, except for the final scraping stage, the harvest, little manpower.

Some Western historians believe that the Chinese may have been the first to develop this technique around A.D. 500. But Chinese historians, who are loath to pass up founder's rights to any invention, lay no claim to this one. The Chinese were not pleased with the salt produced by this technique. Slow evaporation results in coarse salt, and the Chinese have always considered fine-grained salt to be of higher quality.

The idea of successive evaporation ponds seems to have started in the Mediterranean, where coarse salt was valued for salting fish and curing hams. The North African Muslims operating in the early Middle Ages throughout the Mediterranean may have been the first to use such a system, introducing it to Ibiza in the ninth century.

By the tenth century, multiple ponds were being used on the Dalmatian coast, across the Adriatic from Venice. In 965, ponds were built in Cervia, and by the eleventh century, the Venetians had built a pond system.

Venice had intense competition on its little strip of the Adriatic. Close to Venice's Chioggia was Comacchio, where Benedictine monks produced salt. In 932, the Venetians ended that competition by destroying the saltworks at Comacchio. But this served to strengthen the position of the third important saltworks in the area, Cervia, controlled by the archbishop of the nearby no-longer-coastal city of Ravenna.

For a time the two principal salt competitors in the region were the commune of Venice and the archbishop of Ravenna— Chioggia and Cervia. Venice had the advantage because Chioggia was more productive than Cervia. But Chioggia produced *sali minutti*, a fine-grained salt. When Venetians wanted coarser salt, they had to import it. Then, in the thirteenth century, after a series of floods and storms destroyed about a third of the ponds in Chioggia, the Venetians were forced to import even more salt.

That was when the Venetians made an important discovery. More money could be made buying and selling salt than producing it. Beginning in 1281, the government paid merchants a subsidy on salt landed in Venice from other areas. As a result, shipping salt to Venice became so profitable that the same merchants could afford to ship other goods at prices that undersold their competitors. Growing fat on the salt subsidy, Venice merchants could afford to send ships to the eastern Mediterranean, where they picked up valuable cargoes of Indian spices and sold them in western Europe at low prices that their non-Venetian competitors could not afford to offer.

This meant that the Venetian public was paying extremely high prices for salt, but they did not mind expensive salt if they could dominate the spice trade and be leaders in the grain trade. When grain harvests failed in Italy, the Venetian government would use its salt income to subsidize grain imports from other parts of the Mediterranean and thereby corner the Italian grain market.

Unlike the Chinese salt monopoly, the Venetian government never owned salt but simply took a profit from regulating its trade. Enriched by its share of sales on high-priced salt, the salt administration could offer loans to finance other trade. Between the fourteenth and sixteenth centuries, a period when Venice was a leading port for grain and spices, between 30 and 50 percent of the tonnage of imports to Venice was in salt. All salt had to go through government agencies. The Camera Salis issued licenses that told merchants not only how much salt they could export but to where and at what price.

The salt administration also maintained Venice's palatial public buildings and the complex hydraulic system that prevented the metropolis from washing away. The grand and cherished look of Venice, many of its statues and ornaments, were financed by the salt administration.

Venice carefully built its reputation as a reliable supplier, and so contracts with the merchant state were desirable. Venice was able to dictate terms for these contracts. In 1250, when Venice agreed to supply Mantua and Ferrara with salt, the contract stipulated that these cities would not buy salt from anyone else. This became the model for Venetian salt contracts.

As Venice became the salt supplier to more and more countries, it needed more and more salt producers from which to buy it. Merchants financed by the salt administration went farther into the Mediterranean, buying salt from Alexandria, Egypt, to Algeria, to the Crimean peninsula in the Black Sea, to Sardinia, Ibiza, Crete, and Cyprus. Wherever they went, they tried to dominate the supply, control the saltworks, even acquire it if they could.

Producing salt for the Venetian fleet was hard work—moving mud and rocks, clearing and preparing ponds, building the dikes that separated them, carrying heavy sacks of harvested crystal. Often entire families—husband, wife, and children—labored together. They were paid by the amount of salt they harvested.

Venice manipulated markets by controlling production. In the late thirteenth century, wishing to raise the world market price, Venice had all saltworks in Crete destroyed and banned the local production of salt. The Venetians then brought in all the salt needed for local consumption, built stores to sell the imported salt, and paid damages to the owners of saltworks. The policy was designed to control prices and at the same time keep the locals happy. But two centuries later, when a salt fleet en route from Alexandria was lost at sea, the farmers of Crete were in a crisis because salt was so scarce on the island that they could no longer make cheese, which is curdled milk drained and preserved in salt.

In 1473, Venice acquired Cervia, forcing the onetime rival to agree to sell to no one but them. An exception was negotiated for Cervia to continue supplying Bologna in the nearby Po Valley. When Venice's new archrival, Genoa, made the island of Ibiza the largest salt producer in the Mediterranean, the Venetians made Cyprus into the second largest producer. In 1489, Cyprus officially became a Venetian possession.

Aiding its ability to ruthlessly manipulate commerce and control territory, Venice maintained the ships of the merchant fleet as a naval reserve and called them into combat where needed. The Venetian navy patrolled the Adriatic, stopped ships, inspected cargo, and demanded licensing documents to make sure all commercial traffic was conforming with its regulations.

No state had based its economy on salt to the degree Venice had or established as extensive a state salt policy except China.

Possibly this was not entirely a coincidence, since Venetian policy was influenced by one of its best-known families, the Polos.

⌒

IN 1260, NICCOLÒ Polo and his brother Maffeo, both Venice merchants—Venice was by then a city of international merchants—left on a commercial trip to the court of Kublai Khan, a dynamic leader who ruled the Mongols and had just conquered China. They returned in 1269 with letters and messages from Kublai Khan to the pope. The khan asked for more Westerners, intellectuals, and leaders in Christian thought to come to his court and teach them about the West. Two years later the Polo brothers went on a second trip, this time taking with them Niccolòs seventeen-year-old son Marco and two Dominican monks. The Dominicans abandoned the arduous trek, but Marco stayed on with his father and uncle.

If his account is true, no teenager ever went on a better adventure. They traveled the Silk Road across central Asia and the Tarim basin and, four years after leaving Venice, arrived in Shando, or as the poet Samuel Taylor Coleridge called it in his famous poem, Xanadu, the summer capital of Kublai Khan, the emperor of the Mongols. The khan, at least by Marco's account, was not disappointed that the Polo brothers had returned with no greater emissary of Western knowledge than Niccolò's young son. Marco traveled throughout the vast empire the khan had conquered, learned languages, studied cultures, and reported back to Kublai Khan.

Almost twenty-five years later, in 1295, the Polos returned to a Venice, where few, even in their own household, recognized them. Three years after his return, Marco Polo, like other Venetian merchants, was serving in a naval fleet at war with the Venetian rival,

Genoa. He was taken prisoner and supposedly dictated the story of his adventures to a fellow prisoner named Rusticello, a fairly well known author of adventure tales from Pisa.

There are a number of problems with Rusticello. He may have taken great liberties to improve on the story. Whole passages appear to have been borrowed from his previous books, which were imaginary romantic adventures. For example, the arrival of the Polos at the court of Kublai Khan bears a disturbing resemblance to the account of Tristan's arrival in Camelot in Rusticello's book on King Arthur.

From its initial publication in 1300, the Venetians were suspicious. Some questioned if Marco Polo had ever gone to China at all. Why did he write nothing of the Great Wall, about the drinking of tea, about princesses with bound feet? It seemed odd to the few knowledgeable Venetians, and it has seemed suspicious to subsequent scholars that Marco Polo completely missed the fact that China had printing presses in an age when this pivotal invention had not yet been seen in Europe. This omission seemed even more glaring to Venetians 150 years later, once Johannes Gutenberg introduced movable type to Europe, and Venice became a leading printing center.

His book was full of unheard-of details and was missing many of the facts known to other merchants. But later travelers to China were able to verify some of the curious details of Marco Polo's book. And he had been away somewhere for twenty-five years. Polo's account sparked an interest in Chinese trade among many Europeans, including Christopher Columbus, and remained the basis of the Western concept of China until the ninteenth century. His legend has grown.

It is widely accepted that he introduced Italians to pasta. It is true that China at the time, and still today, abounded in fresh

and dried, flat and stuffed pastas. But Marco Polo's book says almost nothing about pasta other than the fact, which he found
very curious, that it was sometimes made from a flour ground
from the fruit of a tree. *Maccheroni,* one of the oldest Italian
words for pasta, appears to be from Neopolitan dialect and was
used before Marco Polo's return. The word is mentioned in a
book from Genoa dated 1279. Most Sicilians are certain that the
first pasta came from their island, introduced by the Muslim
conquerors in the ninth century. The hard durum wheat or
semolina used to make pasta was grown by the ancient Greeks,
who may have made some pasta dishes, and the Romans ate
something similar to lasagna. The word *lasagna* may come from
the ancient Greek *lagana,* meaning "ribbon," or from the ancient
Greek word *lasanon,* which probably would not make the dish
Greek since the word means "chamber pot." The Romans, according to this theory, started using lasanon—presumably not
the same ones but perhaps a similarly shaped vessel—as a pot
for baking a noodle dish.

Marco Polo never mentioned that the Chinese printed paper
money, but it is more significant that he did describe how in
Kain-du salt cakes made with images of the khan stamped on
them were used for money. Among the unexpected details in
Polo's book are many on salt and the Chinese salt administration. Polo described travelers journeying for days to get to hills
where the salt was so pure it could simply be chipped away. He
wrote of the revenue earned by the emperor from the brine
springs in the province of Karazan, how salt was made in Chan-
gli to the profit of both the private and public sector, how Koi-
gan-zu made salt and the emperor derived revenues from it.
Marco Polo seldom mentioned salt without pointing out the
state revenues derived by the emperor.

Marco Polo was a Venetian merchant and may have been genuinely interested in salt and the way it was administered. He also may have decided that, since his readership would be Venice merchants, this would be a subject of great interest to them. But it also could be that whether he had gone to China or not, one of his motivations for writing the book was to encourage the Venetian government to extend its salt administration, especially in possessions around the Mediterranean.

The extent of Marco Polo's influence is difficult to measure, but it is clear that Venice, like the khan, did extend its salt administration and derive great wealth and power from it.

CHAPTER SIX

Two Ports
and the Prosciutto
in Between

W HAT WAS IT about this not especially salty stretch of the Adriatic that made Venetians get into the salt trade, along with the merchants of Cervia, the monks at Comacchio, and the archbishop of Ravenna? It was not so much the sea in their faces as the river at their backs. The Po starts in the Italian Alps and flows straight across the peninsula, spreading into a marshy estuary from Ravenna to Venice. The valley of the Po is an anomaly of the Italian peninsula, so strikingly different that its uniqueness becomes apparent after a moment's glance at a map of Italy. With the Alps to the north and the sylvan mountains of Tuscany to the south, one thick ribbon of rich, rolling green pastures stretches coast to coast along the Po. A haven for agriculture, this has always been the most affluent area in Italy, and today, known as Emilia-Romagna, it still is.

The Romans built a road, the Via Emilia—today it is the eight-lane A-1 superhighway—connecting what became the centers of culture and commerce from Piacenza to Parma to Reggio to Modena to Bologna and on to the Adriatic coast. The agricultural wealth of this region depended on both a port for its goods and a source of salt for its agriculture. By competing for this business, two fiercely commercial competitors at opposite ends of the Po, Genoa on the Mediterranean and Venice on the Adriatic, became two of the greatest ports of the Middle Ages.

On the rich plains of Emilia-Romagna, off of the great Roman road, are the ruins of a Roman city named Veleia. Historians have puzzled over Veleia because the Romans had a clear set of criteria for the sites of their cities and Veleia does not fit them. Not only is it too far from the road, it is on the cold windward side of a mountain. But it has one thing in common with almost every important city in Italy: It is near a source of salt. Veleia was built over underground brine springs, which is why it came to be known as the big salt place, Salsomaggiore.

The earliest record of salt production in Veleia dates from the second century B.C. Like many other saltworks, it was abandoned after the fall of the Roman Empire. Charlemagne, the conquering Holy Roman emperor who, like the Romans before him, had an army that needed salt, started it up again. The name Salso first appears on an 877 document.

In ancient times the brine wells had a huge wheel with slats inside and out for footing. Two men, chained at the neck,

Engraving from the late Middle Ages of a wheel powered by prisoners used to pump brine at Salsomaggiore. State Archives, Parma

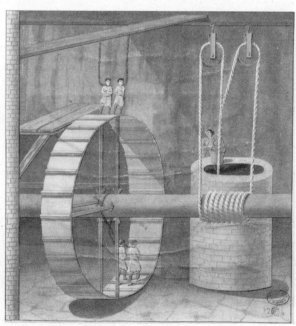

walked inside on the bottom, stepping from slat to slat, and two other men, also chained at the neck, did the same on the outside on top. The wheel turned a shaft that wrapped a rope, which hoisted buckets of brine. The brine was then boiled, which meant that a duke or lord who wished to control the brine wells had to also control a wide area of forest to provide wood for fuel.

Starting in the eleventh century, the Pallovicino family controlled the wells and the region. But in 1318, the city of Parma took over thirty-one Pallovicino wells. The event was considered important enough to be recorded in a fresco in the city palace. He who controlled the brine wells at Salsomaggiore controlled the region, and the takeover of these thirty-one brine wells marked the transfer of power from feudal lord to city government.

IN THE SEVENTH and eighth centuries, before Charlemagne restarted the wells at Salsomaggiore, sailors brought salt from the

A fresco on a wall of the Parma city palace that recorded the city's acquisition of thirty-one wells. The bull is the symbol of Parma. The fresco was destroyed when a tower collapsed on the palace in the seventeenth century. State Archive, Parma

Adriatic to Parma. For this labor they could receive either money or goods, including Parma's most famous salt product, ham—*prosciutto di Parma*.

Parma was a good place to make ham because before the sea air reaches Parma it is caught in the mountain peaks, producing rain and drying out the wind that comes down to the plain. That dry wind is needed for aging the salted leg in a place dry enough to avoid rotting. The drying racks for the hams were always arranged east to west to best use the wind.

Bartolomeo Sacchi, a native of the Po Valley town of Cremona, who became a well-known fifteenth-century author under the pen name Platina, gave blunt and easily followed instructions for testing the quality of a ham:

> Stick a knife into the middle of a ham and smell it. If it smells good, the ham will be good; if bad, it should be thrown away.

The sweet-smelling ham of Parma earned a reputation throughout Italy that was credited not only to the region's dry wind but to the diet of their pigs, a diet which came from the local cheese industry. The Po Valley, where butter is preferred to olive oil, is Italy's only important dairy region. According to Platina, this was more a matter of necessity than taste.

> Almost all who inhabit the northern and western regions use it [butter] instead of fat or oil in certain dishes because they lack oil, in which the warm and mild regions customarily abound. Butter is warm and moist, nourishes the body a good deal and is fattening, yet the stomach is injured by its frequent use.

Notice that Platina listed butter's fattening quality as a virtue, although he was a writer who tended to look for the unhealthy in food including salt, about which he wrote:

"It is not good for the stomach except for arousing the appetite. Its immoderate use also harms the liver, blood, and eyes very much."

And he was not much more sanguine about the pride of his native region, aged cheese.

Fresh cheese is very nourishing, represses the heat of the stomach, and helps those spitting blood, but it is totally harmful to the phlegmatic. Aged cheese is difficult to digest, of little nutriment, not good for stomach or belly, and produces bile, gout, pleurisy, sand grains, and stones. They say a small amount, whatever you want, taken after a meal, when it seals the opening of the stomach, both takes away the squeamishness of fatty dishes and benefits digestion.

The difference between fresh cheese and aged cheese is salt. Italians call the curds that are eaten fresh before they begin to turn sour, *ricotta,* and it is made all over the peninsula in much the same way. But once salt is added, once cheese makers cure their product in brine to prevent spoilage and allow for aging, then each cheese is different.

The origin of cheese is uncertain. It may be as old as the domestication of animals. All that is needed for cheese is milk and salt, and since domesticated animals require salt, that combination is found most everywhere. Just as goats and sheep were domesticated earlier than cattle, it is thought that goat's and sheep's milk cheeses are much older ideas than cow's milk cheese. The habit of carrying liquids in animal skins may have caused the first cheeses since milk coming in contact with an animal skin will soon curdle.

Soon, herders, probably shepherds, found a more sophisticated variation known as rennet. Rennet contains rennin, an enzyme in the stomach of mammals which curdles milk to make it digestible. Usually, rennet is made from the lining of the stomach of an unweaned young animal because unweaned animals have a higher capacity to break down milk. Here, too, salt played a role because these stomach linings were preserved in salt so that rennet from calving season would be usable throughout the year.

The Romans made a tremendous variety of cheeses, with differences not only from one area to another but from one cheese maker to another in the same place and possibly even from one batch to another from the same cheese maker.

Parmesan cheese, now called *Parmigiano-Reggiano* because it is made in the green pastureland between Parma and Reggio, may have had its origins in Roman times, but the earliest surviving record of a Parma cheese that fits the modern description of Parmigiano-Reggiano is from the thirteenth century. It was at this time that marsh areas were drained, irrigation ditches built, and the acreage devoted to rich pastureland greatly expanded. About the same time, standards were established by local cheese makers that have been rigidly followed ever since. Parma cheese earned an international reputation and became a profitable export, which it remains. Giovanni Boccaccio, the fourteenth-century Florentine father of Italian prose, mentions it in *The Decameron*. In the fifteenth century, Platina called it the leading cheese of Italy. Samuel Pepys, the seventeenth-century English diarist, claimed to have saved his from the

London fire by burying it in the backyard. Thomas Jefferson had it shipped to him in Virginia.

In Parma, the production of cheese, ham, butter, salt, and wheat evolved into a perfect symbiotic relationship. The one thing the Parma dairies produced very little of was and still is milk. Just as the Egyptians millennia before had learned that it was more profitable to make salt fish than sell salt, the people of the Po determined that selling dairy products was far more profitable than selling milk.

The local farmers milked their cows in the evening, and this milk sat overnight at the cheese maker's. In the morning, they milked them again. The cheese makers skimmed the cream off the milk from the night before, and the resulting skim milk was mixed with the morning whole milk. The skimmed-off cream was used to make butter.

Heating the mixed milk, they added rennet and a bucket of whey, the leftover liquid after the milk curdled in the cheese making the day before. They then heated the new mixture to a higher temperature, still well below boiling, and left it to rest for forty minutes.

At this point the milk had curdled, leaving an almost clear, protein-rich liquid, and this whey was fed to pigs. It became a requirement of prosciutto di Parma that it be made from pigs that had been fed the whey from Parmesan cheese. Less choice parts of pigs fed on this whey qualified to be sent to the nearby town of Felino, where they were ground up and made into salami. (The word *salami* is derived from the Latin verb *to salt*.)

The cheese makers also mixed whey with whole milk once a week to make fresh ricotta. By tradition ricotta was made on Thursday so that the cheese would be ready for Sunday's traditional *tortelli d'erbette*. *Erbette* literally means "grass," but in Parma it is also the name of a local green similar to Swiss chard. Tortelli d'erbette is a ravioli-like pasta stuffed with ricotta, Parmigiano cheese, erbette, salt, and two spices that were a passion in the thirteenth century and highly profitable cargo for the ships of both Venice and Genoa: black pepper and nutmeg. Tortelli d'erbette was and still is served with nothing but butter and grated Parmigiano cheese.

Before it succumbed to being heavily salted, butter was a rare delicacy. That was especially true in the Po Valley at the southern extreme of butter's range in Europe. In the Parma area, butter was a privilege of the cheese masters—theirs to distribute or sell, generally at high prices. Butter is still sold in the Parmigiano-Reggiano area by cheese masters.

Stuffed pasta in butter sauce worked particular well in this region where the local wheat was soft, different from the rest of Italy, and produced a pasta that, when mixed with eggs, was rich and supple when fresh, but brittle and unworkable when dried. Dried pasta, like olive oil, belonged to the rest of Italy.

Each creamery had a cheese master whose hands reached into the copper vats and ran through the whey with knowing fingers, scooping up and pressing the curds as they were forming. When he said the cheese was done, a cheesecloth was put into the vat, and under his direction the corners of the cloth were lifted, hoisting from the whey more than 180 pounds of drained curd. While the others struggled to suspend the mass in the cheesecloth, only the cheese master was allowed to take the big, flat, two-handed knife and divide the mass in two.

The two cheeses were left one day in cheesecloth and then put in wooden molds. The Latin word for a wooden cheese mold, *forma,* is the root of the Italian word for cheese, *formaggio.* After at least three days, the ninety-pound cheeses were floated in a brine bath turned every day. The aging of cheese is a matter of its slow absorption of salt. It takes two years for the salt to reach the center of a wheel of Parmigiano-Reggiano cheese. After that the cheese begins to dry out. So these cheeses have always had one year of life between when they are sold and when they are considered too hard and dry, even too salty. Platina's admonition about aged cheese may have been a concern about overly aged cheese.

Prosciutto makers used salt from Salsomaggiore, but cheese makers used sea salt supplied either by Venice or by Genoa. In the sixteenth century, the powerful Farnesi family arranged 5,000 mule caravans to carry salt from Genoa's Ligurian coast, known today as the Italian Riviera. Caravans from Genoa carried salt inland to Piacenza, where it was placed on river barges and carried down the Po to Parma. Unlike in Africa or ancient Rome, no single salt route was established. Each caravan had to devise a route based on arrangements with the feudal lords along the way.

Inland cities of the Po Valley such as Parma had their own salt policies and derived revenue from the import of Venetian or Genoese salt, a cost which was passed on to their local consumers. This created a permanent salt contraband trade along the back routes between Genoa, Piacenza, Parma, Reggio, Bologna, and Venice.

In exchange for the salt, the Po Valley traded its salt products: salami, prosciutto, and cheese. It also traded its famous soft wheat for salt. The trade changed with the times. In the eighteenth century, when the Bourbons ruled Parma, they traded their French luxury items for salt, but they also exchanged galley slaves for salt with Genoa, which needed galley slaves for its expanding trade empire. In Parma, a ten-year prison sentence could be reduced to five years as a galley slave on a Genoese ship. But most of these slaves lived only two years, which caused a constant need for replacements.

⌒

In the fifth century b.c., before Genoa was Roman, it was the thriving port of a local people called the Ligurians. It was taken by Rome, by Carthage, by Rome again, by Germanic tribes, by Muslims. Finally, in the twelfth century it became, like Venice, an independent city-state dedicated to commerce.

Genoa bought salt from Hyères near Toulon in French Provence. The name Hyères means "flats" and probably refers to salt flats, because as far back as is known, salt was produced in this place. But in the twelfth century, Genoese merchants turned Hyères into an important producer by building a system of solar evaporation ponds. Genoa's success in Hyères led to the decline of Pisa's Sardinian salt trade. Genoese salt merchants then moved into Sardinia, developed the saltworks of Cagliari, again building a system of evaporation ponds, and made Sardinia one of the largest salt producers in the Mediterranean.

The Genoese also bought salt from Tortosa on the Mediterranean coast of Iberia, south of Barcelona. Tortosa is at the mouth of the Ebro River, which gave it a water connection from Catalonia through Aragon to Basque country—a waterway through the most economically developed parts of the Iberian Peninsula. Tortosa had been a salt producer for the Moors, but by the twelfth century, when Genoa became involved, it was one of the principal suppliers of the port of Barcelona as well as Aragon.

In the mountainous interior of Catalonia, the dukes of Cardona were not happy to see the Genoese selling salt in Barcelona. In 886, a man about whom little is known except possibly his appearance, Wilfredo the Hairy rebuilt an abandoned eighth-century castle on a mountain fifty miles inland from Barcelona. Alone on what was then a distant mountaintop, the highest peak in a rugged, sparsely populated area, he could peer from the thick stone ramparts at his prize possession, the source of his wealth, the next mountain.

This next mountain was striped in pattern and colors so lively, it was almost dizzying to look at it— salmon pink rock with white, taupe, and bloodred stripes. It was all salt. Since salt is soluble in water, elongated facets were cut into the mountain by each rainfall. Inside the mine the pink-striped shafts were ornamented by snow-white crystal stalactites, long dangling tentacles where the salt had sealed over dripping rainwater from fissures above. The salt mountain was by a winding river, a shallow tributary of the

An engraving of Cardona from Voyage pittoresque et historique de l'Espagne (1807–1818). *The castle is on the highest hill, the salt mountain below, by the river. The town where the salt workers lived can be seen in the distance.* Biblioteca de Catalunya

Ebro. Rich green plains and gentle terraced slopes were farmed in the distance, and on the horizon, the snow-crested peaks of the Pyrenees could be seen.

The lords who occupied the castle were the owners of the mountain. A dank brown village of salt workers sprang up on an adjacent mountain. On Thursdays, the salt workers were allowed to take salt for themselves. Starting at least in the sixteenth century, salt workers carved figurines, often religious, from the rock, which has the appearance of pink marble. Soft and soluble, rock salt is easy to carve and even easier to polish.

Even in the area around the salt mountain, all but the top two to three feet of soil is salt, and the white powder leaches to the surface when it rains. There is evidence that people took salt from here as far back as 3500 B.C. Prehistoric stone tools have been found—six-inch-long black rocks with one end serving as a pick and the other as a scraping tool.

The first written record of salt in Cardona is from the Romans, who usually favored sea salt but considered Cardona's rock salt to be of high quality. In the ninth century, the dukes of Cardona, along with the other feudal lords of the Catalan-speaking area, were united under the counts of Barcelona. Catalonia, with its own Latin-based language, became an important commercial power whose territory extended along the Mediterranean coast from north of the Pyrenees to southern Spain.

Cardona was known in medieval Catalonia as an ideal source of salt for making hams and sausages. From the capital, the port of Barcelona, Cardona salt was exported to Europe and became one of the leading rock salts in the Middle Ages. But by the twelfth century, Genoa could bring salt by sea to Barcelona less expensively than the dukes of Cardona could bring it across the fifty-mile land route. As Cardona salt merchants started to lose their Barcelona market, they too began selling their salt to the Genoese.

⌒

AFTER 1250, GENOA went even farther into the Mediterranean, buying salt in the Black Sea, North Africa, Cyprus, Crete, and Ibiza—many of the same saltworks that Venice was trying to dominate. Genoa built Ibiza into the largest salt producer in the region.

Salt was the engine of Genoese trade. With the salt the Genoese bought, they made salami, which was sold in southern Italy for raw silk, which was sold in Lucca for fabrics, which were sold to the silk center of Lyon. Genoa competed with Venice not only for salt but for the other cargoes that were exchanged for salt, such as textiles and spices.

The Genoese were pioneers in maritime insurance, banking, and the use of huge Atlantic-sized ships, which they bought or leased from the Basques, in Mediterranean trade. These ships, with their vast cargo holds, had room for salt on a return voyage. Wherever they went for trade, they made a point of getting control of a saltworks at which to load up for the return trip.

But Venice was winning the competition because of a more cohesive political organization and because of its system of salt subsidies. When this salt competition led to a war in 1378–80, known as the War of Chioggia, Venice's ability to convert its commercial fleet into warships proved decisive. Venice defeated Genoa, its only major competitor for commercial dominance of the Mediterranean.

Yet among those who finally undid the maritime empire of Venice were two Genoese—Cristoforo Colombo and Giovanni Caboto. Neither sailed on behalf of Genoa, and Caboto actually became a Venetian citizen. The beginning of the end came in 1488 when the Portuguese captain Bartolomeu Dias rounded Africa's Cape of Good Hope. In 1492 Columbus, in search of another route to India in the opposite direction, began a series of voyages for Spain, which opened up trans-Atlantic trade carrying new and valuable spices. Then in 1497, Caboto, the Genoese turned Venetian, sailed for England as John Cabot, again looking for a route to India, and told the world about North America and its wealth of codfish. Worst of all, that same year, another Portuguese, Vasco da Gama, sailed around Africa to India and returned to Portugal two years later. Not only were Atlantic ports now needed for trade with the newly found lands, but the Portuguese had opened the way from Atlantic ports to the Indian Ocean and the spice producers. Now the Atlantic, and not the Mediterranean, was the most important body of water for trade.

After the fifteenth century, the Mediterranean ceased to be the center of the Western world, and Venice's location was no longer advantageous. Yet it stubbornly held to its independence and so declined with the Mediterranean.

Genoa succumbed to the new reality, and during Spain's golden age, the Genoese served as the leading bankers and financiers of that expanding Atlantic power. Because of this, Genoa has endured as a commercial center and is today a leading Mediterranean port, though the Mediterranean is no longer a leading sea.

PART TWO

The Glow of Herring and the Scent of Conquest

At the time when Pope Pius VII had to leave Rome, which had been conquered by revolutionary French, the committee of the Chamber of Commerce in London was considering the herring fishery. One member of the committee observed that, since the Pope had been forced to leave Rome, Italy was probably going to become a Protestant country. "Heaven help us," cried another member. "What," responded the first, "would you be upset to see the number of good Protestants increase?" "No," the other answered, "it isn't that, but suppose there are no more Catholics, what shall we do with our herring?"—Alexandre Dumas,
Le grand dictionnaire de cuisine, *1873*

CHAPTER SEVEN

Friday's Salt

B Y THE SEVENTH century A.D., all of western Europe spoke Indo-European languages—languages that stemmed from the Bronze Age Asian invasion of Europe—except for the Basques. In their small mountainous land on the Atlantic coast, partly in what was to become Spain and partly in the future France, Basque culture, language, and laws had survived all the great invasions, including those of the Celts and the Romans.

The Basques were different. One of those differences was that they hunted whales. They were the first commercial whale hunters, ahead of all others by several centuries. The earliest record of commercial whaling is a bill of sale from the year 670 to northern France for forty pots of whale oil from the Basque coastal province of Labourd, which is now in France.

Through the centuries of commercial whaling that would follow, the oil boiled from whale fat would be the most consistently valuable part of the whale. Whalebone was also profitable, especially the hundreds of teeth, which were a particularly durable form of ivory. But, in the Middle Ages, Basque fortunes were made trading the tons of fat and red meat that could be stripped from each whale.

The medieval Catholic Church forbade the eating of meat on religious days, and, in the seventh century, the number of these days was dramatically expanded. The Lenten fast, a custom started in the fourth century, was increased to forty days, and in addition all Fridays, the day of Christ's crucifixion, were included. In all, about half the days of the year became "lean" days, and food prohibitions for these days were strictly enforced. Under English law the penalty for eating meat on Friday was hanging. The law remained on the books until the sixteenth century, when Henry VIII broke with the Vatican.

On lean days sex was forbidden, and eating was to be limited to one meal. Red meat was "hot" and therefore banned because it was associated with sex. However, animals found in water—which included the tails but not the bodies of beavers, sea otters, porpoises, and whales—were deemed cool, and acceptable food for religious days.

For this reason, porpoise is included in most medieval food manuscripts. But the recipes usually call for costly ingredients, indicating that porpoise was not food for the poor. The following English recipe, with its expensive Asian spices, is from a manuscript dated between the fourteenth and fifteenth centuries, though the recipe may be much older.

PURPAYS YN GALENTEYN

Take purpays: do away the skyn; cut hit yn smal lechys [slices] no more than a fynger, or les. Take bred drawen wyth red wyne;

put therto powder of canell [cinnamon], powder of pepyr. Boil
hit; seson hit up with powder of gynger, venegre, & salt.

Fresh whale meat was also for the rich. The great delicacy was
the tongue. Salted tongue of any kind was appreciated, but especially whale tongue. For the poor, there was *craspois,* also called
craspoix, or *grapois.* This was strips of the fattier parts of the
whale, salt-cured like bacon and sometimes called in French *lard
de carême,* which translates as "lent blubber," because it was one
of the principal foods available to the peasantry for lean days on
which other red meats were not allowed. Even after a full day of
cooking, craspoix was said to be tough and hard. It was eaten with
peas, which was the way the rich ate their whale tongue. Nevertheless, Rouen merchants who sold craspoix to the English paid
high tariffs at London Bridge, which suggests this salted whale
blubber was a luxury product in England. This would not be the
last time the food of French peasants was sold as a treat for
wealthy Englishmen.

In 1393, an affluent and elderly Parisian, whose name has
been lost, published a lengthy volume of instructions to his fifteen-year-old bride on the running of a household. The book,
known as *Le mèsnagier de Paris,* offers this recipe:

Craspoix. This is salted whale meat. It should be cut in slices
uncooked and cooked in water like fatback: serve it with peas.

Peas at the time were dried and cooked as beans are today, so
that this dish resembled pork and beans.

On lean days, when the peas are cooked, you have to take
onions that have been cooked in a pot for as long as the peas,

exactly the same way that on meat days, lard is cooked separately in the pot and then peas and stock added. In that same way, on a lean day, at the time the peas are put in a pot on the fire, you should put finely chopped onions and in a separate pot cook the peas. When everything is cooked, fry the onions, put half in the peas and half in the stock—and salt. If that day is during Lent get crapoix and use it the same way that lard is used on meat days.—Le mèsnagier de Paris, *1393*

⌒

BY THE SEVENTH century, the Basques built stone towers on high points of land along their coast. The remains of two still stand. When the lookout in the tower spied a whale, its great shiny black back breaking the surface while spouting vapor, he would shout a series of coded cries that told whalers where and how big the whale was, and how many other whales were nearby. Five oarsmen, a captain, and a harpooner would silently row out, hoping to spear the giant unaware. The Basques, who have always had a reputation for physical strength, made their harpooners legendary—large men of great power who could plunge a spear deep into the back of a sleeping giant.

By the ninth century, when the Basques had a well-established whaling business, an intruder arrived—the Vikings. *Viking* is a term—thought to have its root in the old Norse *vika,* meaning "to go off"—for Scandinavians who left their native land to seek wealth in commerce. They did not have a central location like Genoa or Venice, and their northern home provided them with little to trade. If they had had a source of salt, they might have traded salted meats like the Celts or salted fish like the Phoenicians. But without salt, meat and fish were too perishable, and all the Vikings had to trade were tools made from walrus tusk and reindeer antler. In search of a trading commodity, they raided coastal communities in northern Europe, kidnapped people, and sold them into slavery, which is why they are still remembered for their brutality.

But they were ingenious people, superb shipbuilders, intrepid mariners, and savvy traders. For their captured slaves, they received payment in silver, silks, glassware, and other luxuries that transformed life for the upper classes in Scandinavia. With their fast-sailing ships, they raided the coasts of Britain and France. Starting in 845, these raids turned into campaigns involving large groups. Vikings held territory in the vicinity of the Thames and Loire Rivers, which they used as bases for both raiding and trading at even greater distances. They traded with Russia, Byzantium, and the Middle East. Great European cities, including both London and Paris, paid the Vikings to be left in peace.

The ninth-century Vikings also maintained a base along the Adour River on the northern border of Basque provinces. No records exist of the Vikings teaching shipbuilding to the Basques. But at the time the Vikings built better ships because their hulls were constructed with overlapping planks. And it is known that about that time, the Basques started building their hulls the same way and soon had a reputation as the best shipbuilders in Europe.

With their sturdy, new, long-distance ships equipped with enormous storage holds, the Basques were no longer limited to the whale's winter grounds in their native Bay of Biscay. They loaded their rowboats onto ships and traveled more than 1,000 miles. By 875, only one generation after the arrival of the Vikings in their land, the Basques made the 1,500-mile journey to the Viking's Faroe Islands.

In those cold, distant, northern waters, they discovered something more profitable than whaling: the Atlantic cod. This large bottom feeder preserves unusually well because its white flesh is almost entirely devoid of fat. Fat resists salt and slows the rate at which salt impregnates fish. This is why oily fish, after salting, must be pressed tightly in barrels to be preserved, whereas cod can be simply laid in salt. Also, fatty fish cannot be exposed to air in curing because the fat will become rancid. Cod, along with its relatives including haddock and whiting, can be air-dried before salting, which makes for a particularly effective cure that would be difficult with oily fish such as anchovy or herring.

Had the Vikings told the Basques about cod or perhaps even sold them some? The Vikings knew the cod well from Scandinavian waters. Less than a century after arriving in the Adour, a band of Vikings settled Iceland and then moved on to Greenland and from there, by the year 1000, to Newfoundland. They caught cod as they went and dried it in the arctic air. Realizing that dried cod was a tradable commodity, they soon established drying stations in Iceland to produce the export.

But the Basques had spent centuries surrounded by the Roman Empire, where salted fish was a common food, which is probably why they thought of salt-curing whale meat. Now they started salting cod. The market was enormous. All of the formerly Roman world ate salt fish, and the Basques had a salt fish to sell that, after a day or more of soaking in fresh water, was whiter, leaner, and better, according to many, than the dark, oily, Mediterranean species that had been used before. Being a fatless fish, air-dried and salt-cured, salt cod, stiff as planks of wood, could be stacked on wagons and hauled over roads, even in warm Mediterranean climates. It was better than crapoix and equally affordable, and being a fish, was Church-approved for holy days. For those who desired a more extravagant cuisine, it only needed rich ingredients to dress it up.

Guillaume Tirel, known as Taillevent, was the head chef for King Charles V of France, to whom he introduced cabbage. In the tradition of great French chefs, he worked his way up from a child-hood in a royal kitchen in Normandy, assisting bellows tenders and spit turners rotating enormous roasts, and he hoisted and low-ered the chains holding huge stockpots. Included in his menial jobs as a boy apprentice was desalinating salted meats, regarded as one of the basic skills of an accomplished cook. His nickname Taillevent meant "jib," which is a small, fast, and versatile sail. Four different versions of his manuscript—scrolls of recipes titled *Le viandier*—have been found, all of uncertain dates, but since the working life of Taillevent was from 1330 to 1395, and these were the recipes he used, *Le viandier* is thought to predate *Le mès-nagier de Paris,* making it the oldest known French cookbook.

In *Le viandier,* Taillevent wrote that "Salt cod is eaten with mustard sauce or with melted fresh butter over it." *Le mèsnagier de Paris* borrowed the exact same prescription but added what even today is the best advice on preparing salt cod: "Salt cod that has been too little soaked is too salty; that which has soaked too long is not good. Because of this you must, as soon as you buy it, put it to the test of your teeth, and taste a little bit."

Robert May, a cook for the Royalists in turbulent seventeenth-century England, suggested salt cod be made into a pie.

Being boiled, take it [the salt cod] from its skin and bones, and mince it with some pippins [apples], season it with nutmeg, cin-namon, ginger, pepper, caraway-seed, currans, minced raisons, rose-water, minced lemon peel, sugar, slic't [sliced] dates, white wine, verjuice [sour fruit juice, in this case probably from ap-ples], and butter, fill your pyes, bake them, and ice them.
—*Robert May,* The Accomplisht Cook, *1685*

Though cod is found only in northern waters, salt cod entered the repertoire of most European cuisine, especially in southern Europe where fresh cod was not available. The Catalans became great salt cod enthusiasts and brought it to southern Italy when they took control of Naples in 1443. The following recipe comes from the earliest known cookbook written in Neapolitan dialect.

BACCALÀ AL TEGAME
[PAN-COOKED SALT COD]

Always select the largest cod and the one with black skin, because it is the most salted. Soak it well. Then take a pan, add delicate oil and minced onion, which you will sauté. When it turns dark, add a bit of water, raisins, pine nuts, and minced parsley. Combine all these ingredients and just as they begin to boil, add the cod.

When tomatoes are in season, you can include them in the sauce described above, making sure that you have heated it thoroughly.—*Ippolito Cavalcanti (1787–1860)*, Cucina casereccia in dialetto Napoletano, *Home cooking in Neapolitan dialect*

ALL OF THE fishing nations of northern Europe wanted to participate in the new, rapidly growing, extremely profitable salt cod market. They had the cod but they needed salt, and the Vikings may have been pivotal in solving this problem as well. One of the first Viking bases in the Loire was the island of Noirmoutier. One third of this long thin island, barely detached from the mainland of France at the estuary of the Loire, is a natural tidal swamp, which strong tides periodically flood with a fresh supply of sea-water. The Vikings had long been interested in the use of solar evaporation in making sea salt. Traces of such Viking operations in the seventh century have been found in Normandy. But the northern climate would have made these saltworks unproductive. The climate has too much rainfall and not enough sunlight. It is not known exactly when Noirmoutier, the nearby mainland marshes of Bourgneuf and Guérande, and Ile de Ré, an island about sixty miles to the south, started building systems of artifi-cial ponds, instead of relying on single pond evaporation. As with sharing their shipbuilding skills with the Basques, no record ex-ists of the Vikings teaching artificial pond techniques, but it is known that at the time of their arrival, production greatly in-creased, that the ponds were built sometime in the ninth or tenth

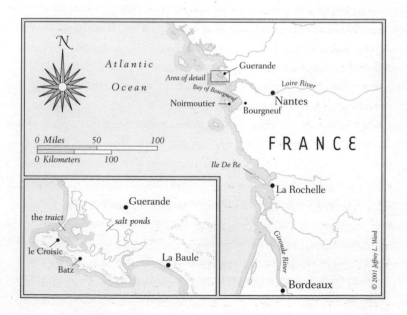

century, and that the Vikings had seen successive artificial pond systems in southern Spain. Since Guérande is in Celtic Brittany, Breton historians with a nationalist streak reject the Viking theory, preferring to believe that Celts originated the idea, which is also possible. It is more certain that the Vikings were the first to trade the salt of this area to the Baltic and other northern nations, establishing one of the most important salt routes of the late Middle Ages and Renaissance.

As Europeans began to recognize that the natural solar evaporation of seawater was the most cost-effective way to produce salt, this area on the southern side of the Brittany peninsula, the Bay of Bourgneuf, became a leading salt center. This bay was the most northerly point in Europe with a climate suited to solar-evaporated sea salt. The bay also had the advantage of being located on the increasingly important Atlantic coast and was connected to a river that could carry the salt inland. Guérande on the north side of the mouth of the Loire River, Bourgneuf on the southern side and the island of Noirmoutier facing them, became major sea salt-producing areas.

⌒

ONE GROUP OF Vikings remained in Iceland, becoming the Icelanders. A second group remained in the Faroe Islands. The main body of Vikings were given lands in the Seine basin in exchange for protecting Paris. They settled into northern France and within a century were speaking a dialect of French and became known as the Normans. Soon the Vikings had vanished.

Meanwhile, Basque ships sailed out with their enormous holds full of salt and returned with them stacked high with cod. They dominated the fast-growing salt cod market, just as they had the whale market, and they used their whale-hunting techniques as a model for cod fishing. They were efficient fishermen, loading huge ships with small rowboats and sailing long distances, launching the rowboats when they reached the cod grounds. This became the standard technique for Europeans to fish cod and was used until the 1950s, when the last few Breton and Portuguese fleets converted to engine power and dragging nets.

Others besides the Basques caught cod in the Middle Ages—the fishermen of the British Isles, Scandinavia, Holland, Brittany, and the French Atlantic—but the Basques brought back huge quantities of salted cod. The Bretons began to suspect that the Basques had found some cod land across the sea. By the early fifteenth century, Icelanders saw Basque ships sailing west past their island.

Did the Basques reach North America before John Cabot's 1497 voyage and the age of exploration? During the fifteenth century, most Atlantic fishing communities believed that they had. But without physical proof, many historians are skeptical, just as they were for many years about the stories of Viking travels to North America. Then in 1961, the remains of eight Viking-built turf houses dating from A.D. 1000 were found in Newfoundland in a place called L'Anse aux Meadows. In 1976, the ruins of a Basque whaling station were discovered on the coast of Labrador. But they dated back only to 1530. Like Marco Polo's journey to China, a pre-Columbian Basque presence in North America seems likely, but it has never been proved.

⌒

FISHERMEN ARE SECRETIVE about good fishing grounds. The Basques kept their secret, and others may have also. Some evidence suggests that a British cod fishing expedition had gone to

North America more than fifteen years before Cabot. The Portuguese believe their fishermen also reached North America before Cabot.

Explorers, on the other hand, were in the business of announcing their discoveries. And they said the cod fishing in the New World was beyond anything Europeans had ever seen. Raimondo di Soncino, the duke of Milan's envoy to London, sent the duke a letter saying that one of Cabot's crew had talked of lowering baskets over the side and scooping up codfish.

After Cabot's voyage, large-scale fishing expeditions to North America were launched from Bristol, St.-Malo on the Brittany peninsula, La Rochelle on the French Atlantic coast, the Spanish port of La Coruña in Celtic Galicia, and the Portuguese fishing ports. Added to this were the many Basque towns that had long been whale and cod ports, including Bayonne, Biarritz, Guéthary, St.-Jean-de-Luz, and Hendaye on the French side, Fuenterrabía, Zarautz, Guetaria, Motrico, Ondarroa, and Bermeo on the Spanish side. On board each of these hundreds of vessels, ranked as a senior officer, was a "master salter" who made difficult decisions about the right amount of salting and drying and how this was done. Both under- and oversalting could ruin a catch.

In the Middle Ages, salt already had a wide variety of industrial applications besides preserving food. It was used to cure leather, to clean chimneys, for soldering pipes, to glaze pottery, and as a medicine for a wide variety of complaints from toothaches, to upset stomachs, to "heaviness of mind." But the explosion in the salt cod industry after Cabot's voyage enormously increased the need for sea salt, which was believed to be the only salt suitable for curing fish.

For the Portuguese, the salt cod trade meant growth years for fishing and salt making. Lisbon was built on a large inlet with a small opening. Aviero, farther up on the marshy shores of the inlet, was an ideal salt-making location. It had been the leading source of Portuguese salt since the tenth century. But with the growing demand, the saltworks at Setúbal, built in a similar inlet just south of the capital, became Portugal's leading supplier. Setúbal's salt earned a reputation throughout Europe for the dryness and whiteness of its large crystals. It was said to be the perfect salt for curing fish or cheese.

Until the sixteenth-century cod boom, La Rochelle had been a minor port because it was not on a river. But suddenly, riverless La Rochelle, because it was an Atlantic port near the Ile de Ré saltworks, became the leading Newfoundland fishing port of Europe. Between Cabot's 1497 voyage and 1550, records show that of 128 fishing expeditions from Europe to Newfoundland, more than half left from La Rochelle, with holds full of salt from Ile de Ré.

The Breton fishing ports also had a salt advantage. Salt was heavily taxed in France, but in order to bring the Celtic duchy of Brittany into the French kingdom, France had offered the peninsula an exemption from the hated *gabelle,* the French salt tax. Though the Breton ports were on the north coast of Brittany, it was only a short distance to the saltworks of Guérande, Noirmoutier, and Bourgneuf.

While northerners had the fish but could not make the salt and southerners had salt but not the cod, the Basques had neither. And yet they managed to get both. By the thirteenth century, they had parleyed their shipbuilding skills into a dependable sea salt supply. They provided the Genoese with their large, well-built ships, and Genoa, in return, gave them access to the saltworks on the island of Ibiza.

England, with its skilled and ambitious fishing fleet and its powerful navy, lacked sea salt. On the Channel coast, sea salt was produced by washing salty sand and evaporating the saltwater over a fire. This method was more costly and less productive than that utilizing the natural solar evaporation of seawater. "For certain uses such as curing fish English white salt and rock salt are not as good as Bay salt which is imported from France," wrote William Brownrigg, a London physician, in his 1748 book, *The Art of Making Common Salt.*

By "Bay salt," he meant solar-evaporated sea salt. The Germans called it *Baysalz.* The reference is to the Bay of Bourgneuf. The coast from Guérande to Il de Ré had become so dominant in salt making that it was synonymous with solar-evaporated sea salt. There were better salts. Northern salts made from boiling peat and southern salts such as that of Setúbal were far whiter, which meant purer. French bay salt was intermittently described as gray, even black or sometimes green. But to northern Europe, it was large-grained, inexpensive, and nearby. An affluent household used bay for curing but more costly white salt for the table. Middle-class homes bought inexpensive bay salt, dissolved it back into brine, and boiled the brine over a fire until crystalized to make a finer salt for serving. *Le mèsnagier de Paris* offers such a recipe for "making white salt."

THE FEW CELTIC places that escaped Romanization are strikingly similar stretches of Atlantic coastline. The low country of southern Brittany with its mud flats in low tide and its marshes full of unexpected canals and ponds is reminiscent of South Wales. South Wales poet Dylan Thomas, describing his homeland—"the water lidded land"—could also have been describing the area of Guérande in the center of a 100,000-acre inland sea with a small opening to the Atlantic. The tides were so powerful that one town, Escoublac, was completely washed to sea in the fourteenth century. After that the salt producers built a seventeen-mile wall separating the sea from the marshland. This wall, which prevents the flooding of 4,400 acres of salt ponds, is still maintained by the salt workers. Here, a salt worker is called a *paludier*, literally a swamp worker.

The tidal area, called the *traict*, has two canals leading to smaller channels leading to an intricate system of large and small saltworks. The paludier let water into his ponds by a series of plugs in small wooden dams. The height of the unplugged holes determined the water level. The paludier held a wooden rake with a long pole and scraped up crystals, piling them on the earthen dikes at the edge of each pond. The piles were left to dry and then hauled away by wheelbarrow. It was a demanding craft because if the clay bottom was disturbed, the salt became black.

In the evenings when a dry wind caused crystals to form on the surface of the water, the women would use long poles with a board on the end to skim the surface and bring in the fleur de sel. This was women's work, because the fleur de sel salt was much lighter and because it was believed the work required a woman's delicate touch, though the dainty work included carrying on their heads baskets of the light salt weighing ninety pounds.

The people of Brittany are Celts, speaking a language derived from the language of Vercingetorix. The paludiers spoke this language until the 1920s. The name Guérande comes from the Breton name Gwenn-Rann, meaning "white country." Other village names include Poull Gwenn, meaning "white port," and Bourc'h Baz, known in French as Le Bourg de Batz, which means "the place coming into view"—because it was on the other side of the salt marsh. Villages of curving streets, lined with one- and two-story stone houses with high-pitched roofs, grew up on the edges of the marsh.

Salt makers carved ponds out of the grassy swamp, where leggy herons and startlingly white egrets waded. It would be easy to get lost in this marsh of tall amber grass with hidden black mud-bottomed waterways. But like mariners at sea, the paludiers could get a bearing from the distant black stone church steeples, especially the Moorish tip of Saint-Guénolé in Le Bourg de Batz, the church named after the patron saint of paludiers. The 180-foot steeple was added to the fifteenth-century church in the 1600s to show navigators the entrance from the marsh to the Loire.

In 1557, 1,200 salt ships from other European ports came to Le Croisic, the rugged port by the opening of the marshy inland sea. Often the number of ships in the harbor far outnumbered the whitewashed stone houses of Le Croisic's few streets. While La Rochelle was becoming France's leading cod port, Le Croisic, between the salt that went out and the goods brought in to trade for salt, became the second most important French Atlantic port after Bordeaux. The British, the Dutch, and the Danish all bought French bay salt. Even the Spanish came to buy bay salt for their fisheries in northern Iberia such as La Coruña.

The Irish, starting in the Middle Ages, traded for salt at Le Croisic. They bought salt for herring, salmon, butter, leather curing, and especially beef and pork. The salt was usually shipped to Cork or Waterford. Their salted beef, the meticulously boned and salted forerunner of what today is known as Irish corned beef, was valued in Europe because it did not spoil. The French shipped it from Brest and other Breton ports to their new and fabulously profitable sugar colonies of the Caribbean—cheap, high-protein, durable slave food. This was later replaced with even cheaper New England salt cod. But the Irish corned beef still traveled far, in part because it was adopted by the British navy—competing with salt cod as a provision.

Irish corned beef became a staple in Pacific islands visited by the British navy, where it is called keg. These islands, especially those of the Hawaiian chain, were well suited for salt making. Hawaiians traditionally made salt for home use by hollowing out a rock to a bowl-like shape and leaving sea water to evaporate in it. They quickly learned to dig evaporation ponds and developed a trade provisioning British, French, and later American ships with salted food such as corned beef, which then became part of their diet as well. Richard Henry Dana, the Harvard graduate who shipped out on the American merchant fleet in the 1830s, in his account of the experience, *Two Years Before the Mast,* which became famous for exposing the appalling conditions on board ship, wrote of the terrible salt beef sailors had to eat in the Pacific. They unkindly labeled it "salt junk."

It was the seventeenth-century English who gave corned beef its name—corns being any kind of small bits, in this case salt crystals. But they did more harm to the name than the Pacific island trade, by canning it in South America. The Irish continued to make it well, and it has remained a festive dish there with cabbage for Christmas, Easter, and St. Patrick's Day, the three leading holidays.

This 1968 recipe by the "Woman Editor" of the *Irish Times* shows the care Irish take with corned beef, and avoids confusion with the lesser English version by calling it spiced beef, which may be closer to the original name.

The following are the ingredients for spicing a six-pound joint:
    3 bay leaves
    1 teaspoon cloves
    6 blades mace
    1 level teaspoon peppercorns
    1 clove garlic
    1 teaspoon allspice
    2 heaped tablespoons brown sugar
    2 heaped teaspoons saltpeter
    1 pound coarse salt

For cooking the meat you will need:
    one six-pound lean boned joint of beef
    three sliced carrots
    a half pint Guinness
    three medium sliced onions
    a bunch of mixed herbs
    one teaspoon each ground cloves and ground allspice

Rub all the dry ingredients together, then pound in the bay leaves and garlic. Stand the meat in a large earthenware or glass dish and rub the spicing mixture thoroughly all over it. This should be done every day for a week, taking the spicing mixture from the bottom of the dish and turning the meat twice. Then wash the meat, and tie it into a convenient shape for cooking.

Sprinkle over about one teaspoon each of mixed allspice and ground cloves, then put it into a large saucepan on a bed of the chopped vegetables. Barely cover with warm water, put the lid on and simmer gently for five hours. During the last hour add the Guinness.

It could be eaten hot or cold, but at Christmas it is usually served cold, in slices. If wanted cold, the meat should be removed from the liquid and pressed between two dishes with a weight on top.—*Theodora Fitsgibbon,* A Taste of Ireland, *1968*

⌒

FOR THE BRITISH, salt was regarded as of strategic importance because salt cod and corned beef became the rations of the British navy. It was the same with the French. In fact, by the fourteenth century, for most of northern Europe the standard procedure to prepare for war was to obtain a large quantity of salt and start salting fish and meat. In 1345 the count of Holland prepared for his campaign against the Frisians by ordering the salting of 7,342 cod caught off the coast. Olaus Magnus, a Swedish bishop, in his 1555 *A Description of the Northern Peoples,* wrote that the provisions necessary to withstand a long siege were herring, eels, bream, and cod—all salted.

The Guérande region specialized in salted fish, including hake, skate, mullet, and eel. In season, May and June, young undersized sardines, known for their delicate flavor, were eaten fresh. The rest of the year, larger sardines were layered in the local salt for twelve days, then washed in seawater and put in barrels. The barrels had holes on the bottom, and on the top was a heavy wooden beam hinged to the wall on one side and weighted with a boulder on the other. The juice was squeezed through the bottom, and every few days another layer of sardines was added until after two weeks the barrel could hold no more.

Other fish were salted, especially during Lent, including mackerel, eel, and salmon. Here is a recipe for whiting—a smaller relative of cod—and one for eel.

Let it die in the salt where you leave it whole for three days and three nights. Then blanch it in scalding water, cut it in slices, cook it in water with green onions. If you want to salt it overnight, clean and gut it. Then cut it in slices; salt by rubbing each slice well with coarse salt.—Le mèsnagier de Paris, *1393*

Take a salt eel and boil it tender, being flayed and trust round with scuers, boil it tender on a soft fire, then broil it brown, and serve it in a clean dish with two or three great onions boil'd whole and tender, and then broil'd brown; serve them on the eel with oyl and mustard in saucers.—*Robert May,* The Accomplisht Cook, *1685*

The incentive of salt cod profits combined with improved artificial pond technology to greatly increase sea salt production, especially in France, but all along the Atlantic as well. And this increase in salt made more fish available. Fishermen, instead of rushing to market with their small catch before it rotted, could stay out for days salting their catch. Expeditions to Newfoundland were out from spring until fall. Salt made it possible to get the rich bounty of northern seas to the poor people of Europe. Salt cod by the bail, along with salted herring by the barrel, are justly credited with having prevented famine in many parts of Europe. The salt intake of Europeans, much of it in the form of salted fish, rose from forty grams a day per person in the sixteenth century to seventy grams in the eighteenth century.

CHAPTER EIGHT

A Nordic Dream

IN SOME PARTS of Sweden it was "a dream porridge," in others a pancake, that was made in silence and heavily salted. The custom was that the girl would eat this salty food and then go to sleep without drinking anything. As she slept, her future husband would come to her in a dream and give her water to quench her thirst.

No data are available on the success rate of Swedish girls using this system to find a mate. But the Swedish dream of salt is well documented. The Swedes had a wealth of herring but nothing with which to salt it.

One of the major commercial uses of salt in the thirteenth and fourteenth centuries was to preserve herring, second only to salt cod, in the European lenten diet. Herring was such a dominant fish in the medieval market that in twelfth-century Paris saltwater fish dealers were called *harengères,* herring sellers.

Herring is a Clupeidae, a member of the same family of small, forked-tailed oily fish with a single dorsal fin as sardines. Anchovies are of a different family but of the same Clupeiforme order as sardines and herring. Even in ancient times, Mediterraneans knew about herring, though they may not have known it as a fresh fish, since it is from northern seas. The Greeks called it *alexium,* from the word *als* or *hals,* as in Hallstatt, meaning "salt." But the people of the Mediterranean world never embraced the salted herring the way they did the salted cod, probably because they had their own clupeiformes. The fact that herring became a hugely successful item of trade in the fourteenth century is directly related to the fact that Atlantic nations, herring producers, were gaining power and controlling markets and commerce in a way they never had before. Antwerp and Amsterdam became leading ports of Europe, far exceeding Genoa and Venice in importance. And just as salt cod became essential to the British and French navies, Dutch ships, both for war and commerce, were provisioned with salt-cured herring.

Herring hide in ocean depths in winter, but in the spring until fall they rise and swim, sometimes thousands of miles, to their coastal spawning grounds. This phenomenon takes place from the Russian and Scandinavian Baltic, across the North Sea, as far south as northern France, and across the Atlantic from Newfoundland to the Chesapeake Bay. Jules Michelet, the poetic nineteenth-century historian, wrote in *La Mer,* "A whole living world has just risen from the depths to the surface, following the call of warmth, desire, and the light."

It is a peculiarity of the English language that while most fish swim in *schools,* herring swim in *shoals,* a word of the same meaning derived from the same Anglo-Saxon root. A herring shoal consists of thousands of fish and, once located, provides an ample catch. But herring feed by gulping in seawater as they swim and filtering out minuscule zooplankton. They will search thousands of miles for these drifting beds of food, which means that a spot that had always been teeming with herring may suddenly one day be devoid of a single one, and they might not return to that spot for years. For the peoples of northern Europe who depended on herring, this could be a cataclysmic event, often blamed on the sins of the village folk. In the Middle Ages, adultery was thought to be a major cause of the herring leaving.

Herring had been a leading source of food for Scandinavians and other populations on the Baltic and North Sea for thousands of years. Archaeologists have found herring bones among 5,000-year-old Danish remains. What really happened in the thirteenth and fourteenth centuries was not so much new ways of salting nor new ways of fishing but an increased supply of salt. This was especially important for herring because the salt had to be readily available for the fishery. Unlike the fat-free cod, herring must be salted within twenty-four hours of being lifted from the sea. This was an almost universally agreed upon and unviable law of herring curing. In 1424 the count of Holland threatened to prosecute any fishermen who cured a herring that had been out of the water for more than twenty-four hours.

There was also an invention of sorts. The standard technique for preserving fish going back to Phoenician times was to gut it, dry it, and pack it in layers with salt. In 1350, Wilhelm Beuckelzon, a fisherman from Zeeland, the fishing center of south Holland—or in other accounts Wilhelm Beucks, a fish merchant in Flanders—started a practice of pickling herring in brine, fresh with no drying at all, and therefore the fish could be cured without the risk of its fat turning rancid from exposure to the air. For centuries, Europe's powers, in their bids for control of the Lowlands, paid homage to Beuckelzon, the inventor of barreled herring. In 1506 Charles V, the Holy Roman Emperor, who was raised in Flanders, visited Beuckelzon's grave to honor his contribution to mankind. But in truth Beuckelzon's invention ranks with Marco Polo's discovery of pasta, or even Columbus's discovery of America, as one of history's more bogus tales.

At the time of Beuckelzon's invention, herring already had been barreled in brine by the Scandinavians, the French, the Flemish, and the English for centuries. Nevertheless, the myth, like many myths, lived on. In 1856 Czar Alexander II of Russia erected a monument to the memory of a fourteenth-century Flemish fisherman named Benkels, who, he said, invented barrel-packed herring and then moved to Finland, thus spreading the idea to Scandinavia. These tributes, even if factually dubious, do speak to the importance northern nations attached to barreled brine-salted herring.

The booming medieval salt fish market was low end—lenten food for poor people. Upper-class people had their fish sped to them fresh, or if they lived too far from the water, had their royal fish ponds and holding tanks, or farmed fish such as carp. But from the sixteenth to the eighteenth centuries an estimated 60 percent of all fish eaten by Europeans was cod, and a significant portion of the remaining 40 percent was herring.

Cured herring had an even lower standing than salted cod, and it was hated by many poor people who had nothing else to eat for holy days. The French way of saying that breeding will tell is *le caque sent toujours le hareng,* the barrel always smells of herring. In Brittany, a rock known to romantics as the Tomb of Almanzor, the legendary tomb of a lover who was drowned at sea, was laughingly known by Breton workers as *tombeau du hareng saur,* the tomb of the pickled herring, because it was where they ate their lunch.

But despite its low standing, fortunes could be made on furnishing the poor with herring. The herring was plentiful. Access to salt was the only limitation.

⌒

EVAPORATING SEAWATER OVER a fire was slow and costly, but northerners developed techniques to produce salt in gray, rainy climates. In northern Holland and southern Denmark, peat salt was made by burning peat that was impregnated with seawater. This ocean-soaked peat, known in Dutch as *zelle,* was dug from the tidal flats off the coast. The Dutch sometimes built temporary dikes to seal off an area while zelle was being harvested. It would be loaded on boats and carried to the mainland.

In the Middle Ages, Zeeland, the ocean-pocked estuary of the Schelde River in southern Holland, was a center for peat salt. Porters would carry the sea-logged peat to huts, where it was dried and burned. All that would remain would be ashes and salt. Saltwater was added, and it would absorb the salt in a brine and leave behind the ashes. The brine was then evaporated. When badly made, this produced an impure product known as black salt. But it could also produce a very white, fine-grained salt if the peat was not mixed with soil and if an unscrupulous salt maker had not bulked up his crystals by deliberately adding white ashes. Good-quality peat salt from the Lowlands was highly valued for herring but was expensive and produced only in small quantities.

By the mid–thirteenth century, good-quality zelle was becoming hard to find and very valuable, which made peat salt even more expensive. The greedy would pilfer dikes, the earthworks that kept out the sea, for peat. The one sacrosanct law of Dutch society, a nation living at sea level and below, is to preserve the dikes. Salt makers began to be seen as a threat to this critical first

Salt making from Olaus Magnus's 1555 A Description of the Northern Peoples. Kungliga Biblioteket

line of national defense, and laws were passed heavily taxing zelle, then fining anyone who dug peat within the Zeeland dikes, and, in time, repressing the salt industry.

Some sea salt was produced on the southern shore of England, but only in unusually dry, sunny summers. On the Danish island of Laesø, in the body of water known as the Kattegat, which lies between Denmark and Sweden, sea salt was produced by evaporating ocean water to a denser brine and then boiling it. The Finnish, too, made salt by this process, boiling down arctic seawater near the current Russian town of Murmansk. The salt was mostly used for the productive salmon fishery of the area, but some of it was shipped by cart to Finland and Russia. The Norwegians used a similar process. Though this salt was expensive, the demand made it economically feasible. Oslo was actually a salt trading center.

Olaus Magnus described how Norwegians improved upon this arduous sea salt process by pumping saltier water from sea depths by means of piping made from hollowed tree trunks. The same practice was carried out in Sweden until the eighteenth century. The process destroyed a great deal of forest, the source of fuel and piping, to produce only a small amount of salt.

Sweden hoped to acquire an island in the Caribbean from which to produce salt, but when it finally got one, St.-Barthélemy, the amount of salt produced and shipped back to Sweden was barely enough to cure the quantity of herring destined for the island as slave food.

The shortage of salt in the North was frustrating because of all the world's oceans, the cold subarctic seas have the densest schools and the greatest variety of species. Magnus wrote:

> Herring can be purchased very cheaply for the supply is copious. They present themselves in such large numbers off shore that they not only burst the fishermen's nets, but, when they arrive in their shoals, an axe or halberd thrust into their midst sticks firmly upright.—*Olaus Magnus,* A Description of the Northern Peoples, *1555*

Landing herring from Olaus Magnus's 1555 A Description of the Northern Peoples. Kungliga Biblioteket

The abundance of herring, combined with their method of extracting food from swallowing seawater and the fact that they appeared to die instantly when taken from the water, led some medieval observers of natural phenomena to conclude that herring was a unique species of fish whose only nourishment was the seawater itself. Adding to their mystique, herring seem to let out a cry when they die, a high-pitched hiss, which is probably air escaping the swimming bladder.

The small fish were also noted by maritime people for a phenomenon known as herring lightning, which occurred because the shoals were so dense that they reflected light.

In the sea at night its eyes shine like lamps, and, what is more, when these fish are moving rapidly and the huge shoal turns back on itself, they resemble flashes of lightning in the churning water.—*Olaus Magnus,* A Description of the Northern Peoples, *1555*

A NUMBER OF ways were found to preserve herring with small quantities of salt. The Dutch had their *groene haringen,* green herring, sometimes called new herring, which was gutted on board ship in early spring or late fall—before or after spawning. The herring 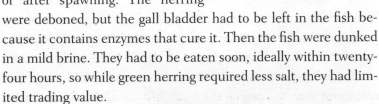 were deboned, but the gall bladder had to be left in the fish because it contains enzymes that cure it. Then the fish were dunked in a mild brine. They had to be eaten soon, ideally within twenty-four hours, so while green herring required less salt, they had limited trading value.

For both meat and fish, smoking was a northern solution to a lack of salt. Salt is needed for smoking but in smaller quantities, because the smoking aids in conservation. The origin of smoking is unknown. The Romans smoked cheeses and ate Westphalian ham, which was smoked. It is not known when the first fish was smoked. In the 1960s, a Polish archaeologist found a fish smoking station in the area of Znin, which he dated from between the eighth and tenth centuries. The Celts and Germans did not lack salt, yet they smoked their hams because cold winters forced food to be enclosed in fire-warmed rooms.

Smoked foods almost always carry with them legends about their having been created by accident—usually the peasant hung the food too close to the fire, and then, imagine his surprise the next morning when . . .

Red herring, a famous export from the East Anglia region of England along the North Sea, is soaked in a brine of salt and saltpeter and then smoked over oak and turf. The discovery of red herring was described by a native East Anglian, Thomas Nash, in 1567. He claimed that it came about when a Yarmouth fisherman with an unusually large catch hung the surplus herring on a rafter, and by chance the room had a particularly smoky fire. Imagine his surprise the next day when the white-fleshed fish had turned "red as a lobster."

Finnan haddie, a haddock soaked in brine and then smoked over peat and sawdust, was originally called Findon haddocks because it was made in the Scottish North Sea town of Findon, near Aberdeen. It was not commercialized until the mid–eighteenth century, though it may have been a household product for a long time before that. Despite the relatively recent date, it is commonly said that finnan haddie too was originally made accidentally by fishermen hanging their salt fish too close to the smoky peat fire in their cabin.

By the sixteenth century, if not earlier, on the Swedish coast of the Gulf of Bothnia, the body of water between Sweden and Finland, a light cure was devised for Baltic herring, and these pickled fish became known as *surströmming*. The Baltic Sea, a less salty body of water than the North Sea, has leaner and smaller herring than the Atlantic and North Sea herring eaten by the British and the Dutch. In Sweden, which has both a North Sea and a Baltic coast, the fish are known by completely different names. A Baltic herring is called a *strömming,* and a North Sea herring is a *sill*. A number of Baltic languages make this distinction. Russians speak of a Baltic *salaka* and an Atlantic *sel'd'*.

A story persists in Sweden that surströmming was discovered by accident by Swedes trying to save on salt. Surströmming was a basic ration of the Swedish army in the seventeenth century during the fifty years of sporadic armed conflict that is known as the Thirty Years War. It is still regulated by a medieval royal ordinance and must be made from herring caught in April and May just before spawning. The head and entrails are removed, but the roe is kept in the herring, which is put into light brine in barrels holding 200 pounds of fish. The fish are left to ferment in the barrels for ten to twelve weeks at a temperature between fifty-four and sixty-four degrees Fahrenheit. The third Thursday in August, the producers are allowed to put the fish on the market.

Originally it was taken from the barrel, but in modern times it is canned in July. By eating time in September, the can is bulging on the top and bottom and looks ready to explode. As the can is opened, the family stands around it to get the first fumes. Nowadays some of the younger members flee the room. The can opener digs in, and a white milky brine fizzes out, bubbling like fermented cider and smelling like a blend of Parmesan cheese and the bilge water from an ancient fishing vessel.

These potent little fish have always been shrouded in controversy because, like Roman garum, they flirtatiously hover between fermented and rotten. Like garum, through, surströmming
is in truth fermented and not rotten, because the brine the fish is
dipped in is sufficient to prevent putrification until the fermentation process takes over. If done properly, surströmming has a
strong flavor, one revered by aficionados of cured fish and loathed
by the less initiated.

To eat surströmming, the bloated, bluish-white, little headless
fish is slit in the belly and the roe removed. None but the brave
eat the roe. The splayed fish is mashed hard on the spine with a
fork and turned over. The bones can then be easily lifted off. The
wine-colored fermented flesh inside is then placed on a buttered
krisp, a Swedish cracker, with mashed potatoes. Swedes use a
small long yellow fingerling potato with a floury texture—a breed
designed to survive the northern winter. In the north of Sweden,
onions are added, but in the south this is regarded as an unnecessary distraction. Once properly blended with all these tastes
and textures, the fish is surprisingly pleasant. The only remaining
problem is how to get the smell out of the house, a lingering odor
that suggests the question: How could such a thing possibly have
been eaten? In recent years a Swedish company tried to export
surströmming to the United States, but the U.S. government refused it entry on the grounds that it was rotten.

THE MORE USUAL way of preserving fish required a great deal
of salt. Herring salting was described by Simon Smith, an agent
for the British government, in 1641. As soon as the herring were
taken from the nets, they were passed to "grippers," who gutted
them and mixed them with dry salt crystals and packed them in a

barrel. The barrels were then left for a day to draw out the herring juice and dissolve most of the salt. Then more salt was added and the barrel closed. According to Smith, the brine had to be dense enough for the herring to float. A barrel containing 500 to 600 herring would require fifty-five pints of salt.

The salt shortage of the northern fisheries was solved by a commercial group that organized both herring and salt trades. Between 1250 and 1350, a grouping of small associations in northern German cities formed. Known as the Hanseatic League, from the Middle High German word *Hanse*, meaning "fellowship," these associations pooled their resources to form more powerful groups to act in their commercial interests. They stopped piracy in the Baltic, initiated quality control on traded items, established commercial laws, provided reliable nautical charts, and built lighthouses and other aids to navigation.

Before the Hanseatics gained control of the northern herring trade, peat salt was often laced with ashes, and inferior, even rotten herring was commonly sold. *Le mèsnagier de Paris* gave this advice: "Good brine cured herring can be recognized because it is lean but with a thick back, round and green, whereas the bad ones are fat and yellow or the backs are flat and dry."

The fine, round-backed ones would be nicely displayed on the top of the barrel, and then only a few layers down lay the flat, dry backs. The Hanseatics guaranteed that an entire barrel was of quality. Those caught placing bad herring in the bottom of a barrel were heavily fined and forced to return the payment they received. The inferior fish were burned and not thrown back into the sea for fear other fish would eat them and become tainted.

By the fourteenth century, the Hanseatics controlled the mouths of all the northerly flowing rivers of central Europe from the Rhine to the Vistula. They had organizations in Iceland, in London, and as far south as the Ukraine and even in Venice. This gave them the ability to buy salt from numerous sources to supply the northern countries. In the early fourteenth century, the Hanseatics, realizing the low prices and light tax on Portuguese salt more than made up for the cost of transporting it a longer distance, imported Setúbal white salt to trade in the Danish and Dutch fisheries. In the year 1452 alone, 200 Hanseatic ships stopped in Le Croisic to load Guérande salt for the Baltic.

In the fourteenth and fifteenth centuries, Fasterbö and Skanör in southern Sweden became major herring producers. They imported salt from the Hanseatic German port of Lübeck and exported their cured herring back to Lübeck to be marketed throughout Europe. All of this trade was carried on Hanseatic ships. At the height of their power in the fifteenth century, the Hanseatics were believed to have had at their command 40,000 vessels and 300,000 men.

For a time, the Hanseatics were well appreciated as honorable merchants who ensured quality and fought against unscrupulous practices. They were known as Easterlings because they came from the east, and this is the origin of the word *sterling*, which meant "of assured value." The Hanseatics are still honored by a street name—Esterlines Street—in the medieval port district of the Basque city of San Sebastián.

But in time they were seen as ruthless aggressors who wanted to monopolize all economic activity, and the merchant class rebelled against them. To control herring and salt was to control northern economies. In 1360, the Danes went to war with the Hanseatics over control of herring and lost. By 1403, when the Hanseatic League gained complete control of Bergen, Norway, it had achieved a monopoly on northern European production of herring and salt but not without constant warfare with rebellious Baltic states. In 1406, the Hanseatics caught ninety-six British fishermen off Bergen, tied their hands and feet, and threw them overboard.

Baltic herring started to vanish—perhaps too much adultery was being committed in Baltic villages—and the North Sea catches became larger than ever. Suddenly the strömming had vanished and the sill abounded. This strengthened the English and the Dutch and weakened the Hanseatic League. Slowly the British and Dutch gained enough economic and military might to overwhelm the cartel. This was especially true after colonization gave them North American fisheries.

But once the Hanseatic League began to fade, the Dutch and British were still in competition. Their herring fisheries, which became the European leaders, faced each other across the North Sea in Brielle on the Dutch side and Yarmouth on the English side.

The seasonal arrival of the herring shoals became essential to the economies of both England and Holland. In medieval England, every spring, lookouts were posted along the important seaward points of eastern Britain to spot the arrival of the herring. The lookouts would point with a stick to indicate the direction the shoal was swimming from the first point off Crane Head in the Shetlands in early June until they reached Yarmouth in September. In Yarmouth, as early as the fourteenth century, the annual fair marking the end of the herring season, held from September 29 until November 10, attracted herring merchants from the rest of Europe.

Like the Venetian salt fleet, the huge Dutch herring fleet was trained as an armed naval force that fought numerous wars in Europe and the Caribbean against the British professional navy. Finally, in 1652, the British navy destroyed the Dutch herring fleet. In time, the Dutch made peace with the British. England got a Dutch king. But this still left the French, who had their own herring fleets and every intention of controlling salt and being a world power.

CHAPTER NINE

∼

A Well-Salted Hexagon

I N A 1961 speech, Charles de Gaulle, explaining the un-governable character of the French nation, said, "Nobody can easily bring together a nation that has 265 kinds of cheese." The reason for the variety is that, given its limited area, the amalgamation that became France had a remarkable diversity of climates, topography, and cultures. The nation was slowly constructed from feudal kingdoms. It included Burgundians and Provençals, Germanic-speaking Alsatians, Celtic-speaking Bretons, Basques, and Catalans. The Hexagon, as the French would come to call it, bordered the Lowlands, the Rhine, the Alps, the Mediterranean, the Pyrenees, the Atlantic, and the English Channel, which the French have never called English but simply La Manche, the sleeve, a word that refers to its long and narrow shape. The Hexagon offered a wealth of salt: rock salt, brine springs, and both Mediterranean and Atlantic sea salt.

The royal tables of the diverse medieval and Renaissance

French kingdoms were set with huge, ornate *nefs*, ships, in this case jeweled vessels holding salt. A nef was both a saltcellar and a symbol of the "ship of state." Salt symbolized both health and preservation. Its message was that the ruler's health was the stability of the nation.

In 1378, Charles V of France hosted a famous dinner that posed the awkward question of where to place the nef. Should it be in front of him or by his guest Charles IV, the Prague-born Holy Roman Emperor? And what about the emperor's son who was also joining them, King Wenceslaus of Germany, who would become emperor after his father's death later that same year? It was decided that the table had to be set with three large nefs, one for each of the three monarchs.

Richard II, the fourteenth-century British monarch whose unpopularity was attributed to both his gaudy extravagance and his lackluster pursuit of the Hundred Years War with France, had a nef on his table with figures of eight tiny men on the ship deck hoisting the flags of France. The unusual nef had no shortage of admirers, since Richard employed 2,000 chefs and was said to have entertained 10,000 visitors daily, most of whom stayed for dinner.

In the fifteenth century, Jean, duc de Berry, featured on his banquet table a gold ship that held not only salt but pepper, as well as, according to some accounts, powdered unicorn horn. Since it is doubtful that anyone has ever seen a unicorn, the powder may have been from the tusk of a narwhal, a single-horned relative of the whale. Unicorn horn was believed to be a poison antidote, which many monarchs wanted to have close-by at mealtime. Some nefs contained "serpent's tongue," which was actually shark's tooth, for the same purpose. The compartments in nefs were frequently locked.

Elaborate saltcellars in all forms, not only ships, were popular. In addition to his nef, in 1415 the duc de Berry, a notable patron of the arts, received from the artist Paul de Limbourg an agate saltcellar with gold lid and a sapphire knob with four pearls.

In the sixteenth century, when things Italian were especially fashionable, Benvenuto Cellini, the Florentine high-Renaissance sculptor and goldsmith, made a saltcellar for King François I of France, perennial war maker and insatiable art enthusiast. The dish of salt was held between the figure of Neptune, god of the sea, and the earth goddess—salt between its two sources, sea and earth. By Neptune's knee was a temple with a tiny drawer for pepper.

In addition to an elaborate saltcellar, referred to as the Great Salt, lesser saltcellars would be removed from the table and oth-

Benvenuto Cellini's saltcellar for François I. Kuntshistorisches Museum, Vienna

ers brought out with changing courses. The Great Salt stayed by the master or host or most honored guest throughout the meal.

It was considered rude, sometimes even unlucky, to touch salt with the fingers. Salt was taken from the cellar on the tip of a knife and a small pile put on the diner's plate. Some medieval and Renaissance plates had a small depression for salt.

Placing salt on the table was a rich man's luxury, but all classes ate salted foods. In 1268, the *Livre des métiers,* the Book of Trades, which listed the rules of the cooking profession, said that cooked meat could be kept for only three days unless it had been salted. *Le mèsnagier de Paris* gave recipes for not only salted whale, but also beef, mutton, venison, coot (an aquatic bird), goose, hare, and a great number of pork products. Although salting was often done in the home, it was usually not women's work. The medieval French, like the Chinese, believed that the presence of women could be destructive to fermentation. In France, a menstruating woman is said to be *en salaison,* curing in salt. It was dangerous to have a woman in a room full of fermenting food when she herself was in fermentation. "It will spoil the lard," people would say.

Originally, salting was a way to keep food through the winter, but by the Middle Ages such foods were eaten year-round.

In June and July pieces of salted beef and mutton should be well cooked in water with green onions after having rested in salt from morning to evening or for a day or more.—Le Mèsnagier de Paris, *1393*

A food that typifies the French love of salted foods is the *choucroute* of Alsace and Lorraine. Alsace, known as Elsass in German, was part of the Holy Roman Empire, and France did not add it to its Hexagon until 1697. The Alsatian language is a dialect of German. Choucroute appears to have evolved from German sauerkraut. But the French, having a resistance to acknowledging German origins in their culture, argue that the Chinese salted cabbage, and the Tartars made it, and, always the favorite French source for foreign food, Catherine de Médicis might have introduced it. Catherine was a sixteenth-century Florentine who married the future Henri II of France and moved to his country with many Italian food ideas.

In a popular French legend, superstar Sarah Bernhardt went to a Chinese restaurant in Paris and ordered choucroute. The waiter fetched the maître d', who with a suggestion of indignation informed the actress, "This is a Chinese restaurant, Madame."

"Mais oui, Monsieur," the actress supposedly replied. "Choucroute is a Chinese invention."

The Chinese may not have invented it. Scientists have found evidence of early hunters curing a leaf that resembled cabbage. But the Chinese have been pickling vegetables for millennia, and cabbage was one of the first vegetables used.

The Romans made sauerkraut and were great cabbage enthusiasts. Cato suggested that women would live long, healthy lives if they washed their genitals in the urine of a cabbage eater. He was listened to on health matters, since in an age of short lives and high infant mortality he lived to be over eighty and claimed to have fathered twenty-eight sons, all of which he credited to eating cabbage with salt and vinegar.

On the other side, Platina, from fifteenth-century Cremona, had warned against it:

It is agreed that cabbage is of a warm and dry nature and for this reason increases black bile, generates bad dreams, is not very nourishing, harms the stomach a little and the head and eyes very much, on account of its gas, and dims the vision.

The Alsatian word for choucroute, *surkrut*, resembles the German, *sauerkraut*. Both words have the same meaning: sour or pickled grass. The German princess Palatine, sister-in-law to Louis XIV, claimed to have introduced the dish to the court at Versailles. She wrote back to her sister in Germany, "I have also made Westphalian-type raw hams fashionable here. Everyone eats them now, and they also eat many of our German foods—sauerkraut, and sugared cabbage, as well as cabbage with fat bacon, but it is hard to get it of good quality." She sent to Germany for cabbage seeds but still complained that the vegetable did not grow well in sandy French soil.

The closest to German soil would be the west bank of the Rhine—Alsace. The word *Alsace*, with the root *als,* may have originally meant "land of salt." The rock salt of Alsace does not have a high concentration of sodium chloride but has a considerable concentration of potassium chloride, known as potash, and in modern times the Alsatian habit of mining the potash for fertilizer and dumping the sodium chloride in the Rhine has become a major environmental issue.

Until 1766, Lorraine was the independent kingdom of Lotharingia, named after the ninth-century king Lothair. Long before France acquired it, Lorraine was already famous for the richness of its brine springs, which have denser brine than in most of Germany and have been exploited since prehistoric times. In the Seille Valley of Lorraine, salt has been produced since the time of the Celts. The Seille, whose name means "salty," is a tributary of the Moselle. The Celtic salt mine had been abandoned, but in the tenth century, Lotharingians began boiling brine with wood fires. Salt could be moved along the Moselle to Alsace, Germany, and Switzerland. Choucroute, surkrut, and sauerkraut were all made with Lorraine salt.

Surkrut was a dish for special occasions—weddings and state banquets. By the sixteenth century, a trade existed in Alsace known as a *surkrutschneider*. Literally sauerkraut tailors, surkrutschneiders chopped cabbage and salted it in barrels with anise seeds, bay leaves, elderberries, fennel, horseradish, savory, cloves, cumin, and other herbs and spices. Each surkrutschneider had his own secret recipe.

By the early eighteenth century, the French had their own word for surkrut: *sorcrotes*. In 1767, encyclopedist and philosopher Denis Diderot mentioned *saucroute* in a letter, and finally, in 1786, on the eve of the French Revolution, the word *choucroute* first appeared. By then, choucroute was generally served mixed with or as a bed underneath other salted foods, and the dish was called *choucroute garnie*. Originally it was served with salted fish, especially herring. But gradually salted fish was replaced by salted meats—an assortment of sausages and cured cuts of pork piled festively on a large platter of seasoned, salt-cured cabbage.

Like wine, salt, and salted meats, choucroute was an important international trading commodity for Alsace.

Sauerkraut made its notable debut off the continent in 1753, when an English doctor informed the admiralty that it prevented scurvy. Many medieval Europeans had heeded Cato's words on the health benefit of cabbage, fashioning plasters and cough medicines from the leaves. The healing of the Holy Roman Emperor Maximilian II by application of cabbage plasters in 1569 was widely publicized.

Once again, cabbage was medicine. The British navy set up "sauerkraut stores" in British ports so that all Royal Navy vessels could set sail provisioned with sauerkraut. Captain James Cook had it served to his crew with every meal. At the same time, across the channel in Paris, it remained banquet food for the royal court. Marie Antoinette, whose father was from the house of Lorraine, championed choucroute at court. This classic early-twentieth-century recipe was little changed from that time.

You can buy choucroute ready-made at the Charcuterie or at a prepared-food shop; but in the countryside this is difficult to find; we are giving the recipe in the simplest way possible.

Take a round, well-shaped white cabbage, clean it, pulling off the green or wilted leaves, split the cabbage in quarters and remove the thick sides that form the heart. Then cut the cabbage in slices thick as a straw and prepare the following brine:

Take a small barrel that once contained white wine, clean it thoroughly and cover the bottom with a layer of coarse salt; then put on top of this a layer of cabbage cut into little strips; sprinkle with juniper berries and here and there peppercorns, press the layer in well but without letting it break up; add a new layer of salt, a bed of cabbage, juniper berries and pepper and continue taking care to press well.

It takes about two pounds of salt for a dozen cabbages. The barrel should be only three quarters filled, cover the choucroute with a piece of loosely-woven linen, then a wooden lid that fits completely inside the barrel. Put a 30 kilo [66 pound] weight on the lid. (If you don't have a weight use a rock or a paving stone.) Once fermentation begins, which happens after a short while, the lid drops down and becomes covered with water formed by the salt. You remove this water but take care to leave a little on the lid.

You can use the choucroute at the end of a month, but once you take some, you have to take care to wash the cloth and the lid and replace them, adding a little fresh water on the top to replace what has been taken.

The fermentation gives the choucroute a bad smell, but don't worry about it, because you wash the choucroute with a little water before serving and the bad taste vanishes.

Choucroute garnie: Wash the choucroute, changing the water several times, and squeeze it well in your hands; when it is drained so that no more water is in it, prepare a casserole, placing a piece of lard in the bottom (the fat side should touch the dish). Place on it a bed of choucroute not too tightly packed, salt, pepper, juniper berries, a bit of fat from a roast, a small slice of *lard maigre* [a baconlike cut of pork from the chest], a small sausage, and half a raw *cervelas* [a stumpy all-pork sausage, usually seasoned with garlic. The name comes from pork brains, *cervelle,* which are rarely included anymore]. Put in another layer of choucroute, salt, pepper, juniper berries, roast fat, lard maigre, sausage, cervelas, and continue in this way until there is no more choucroute. Moisten the entire thing with a bottle of white wine and two glasses of stock. Cover and let cook for five hours over a low heat.

Finally, remove the fat from the top and press hard on the choucroute with a spoon, then place a platter over the casserole and turn it upside down so that your choucroute comes out shaped like a pâté.

For two or three people, you can use two pounds of choucroute. It is better reheated the next day.—*Tante Marie,* La veritable cuisine de famille (*Real family cooking*), 1926

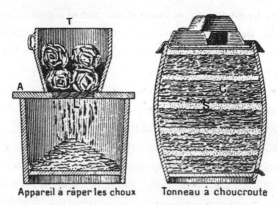

Appareil à râper les choux   Tonneau à choucroute

A diagram showing how to make choucroute from the 1938 edition of
Larousse Gastronomique *by Prosper Montagné.*

IN THE TIME of Pliny, a Roman legionnaire named Peccaius established a sea-salt pond in the estuary of the Rhône to raise
money to pay the salaries of the enormous Roman army fighting
in Gaul. The marshy area, including the swamp known as the Camargue, was well suited to salt making. Located both on the
Mediterranean and on a river that led into Gaul and later France,
the Italians recognize the Rhône estuary as an ideal location for
their salt making. A short distance from where the Genoese had
their saline in Hyères, other Italians, especially the Tuscans, invested in saltworks in the estuary of the Rhône.

In the thirteenth century, a group of religious extremists based
in the town of Albi and known as the Albigensians, launched a series of crusades to cleanse the region of "heretics." Asked how to
recognize a heretic from a true believer, one Albigensian leader,
according to legend, said, "Kill them all. God knows his own."
The chaos that ensued from this approach is known as the Albigensian Wars. In 1229, Louis IX, the fifteen-year-old king of

France, concluded a treaty to end the French campaign against the Albigensians, in which the Rhône estuary was ceded to the French Crown.

This gave France its Mediterranean coast, and in 1246 Louis established the first French Mediterranean port, a walled city named Aigues-Mortes, which means "dead waters." These dead waters lay beyond the massive ramparts that enveloped the city, in a vast expanse of salt evaporation ponds built out into the Mediterranean. Louis wanted salt revenue to finance his dream of leading a Crusade to the Middle East, which he did two years later. He captured an Egyptian port before being defeated and taken prisoner. For this he has been ever after known in French history as Saint Louis. When he finally returned to France in 1254, his saltworks, milky ponds where tall pink flamingos waded, were still producing salt and state revenue.

In 1290, the Crown bought nearby Peccais, site of the Roman works, and the two became the third largest producer of salt in the Mediterranean Sea, after Ibiza and Cyprus. This idea of Saint Louis, for Mediterranean saltworks to be controlled for royal revenue, would one day grow into what would be remembered as one of the greatest disasters of French royal administration.

⌒

THE MEDITERRANEAN SALTWORKS shipped their product up the Rhône as far as Lyon. The salt was also carried on land routes over the mountains of Provence to nearer destinations such as Roquefort-sur-Soulzon.

It is the presence of salt throughout France, along with either cows, goats, or sheep, that has made it the notoriously ungovernable land of 265 kinds of cheese. French cheese makers were trying to be neither difficult nor original. They were all trying to preserve milk in salt so they could have a way of keeping it as a food supply. But with different traditions and climates, the salted curds came out 265 different ways. At one time there were probably far more variations than that.

The cheese made in the Aveyron, a mountainous area of dramatic rock outcropping and thin topsoil, is as old as its famous salt source. Pliny praised a cheese from these mountains above the Mediterranean coast that was probably a forerunner of the now famous Roquefort. According to a widely believed, though not well-documented, legend, Charlemagne passed through the area after his disastrous Spanish campaign of 778. The monks of the nearby monastery of St.-Gall served the emperor Roquefort cheese, and he immediately busied himself cutting out the moldy blue parts, which he found disgusting. The monks convinced him that the blue was the best part of the cheese—an effort for which they were rewarded with the costly task of providing him with two wheels of Roquefort a year until his death in 814.

The ancient cheese was and still is produced from milk of the sheep that graze on the rugged slopes of a hidden mountain area named after its largest town, the village of St.-Affrique. Although St.-Affrique has a humid climate, its soil cannot produce crops because the porous limestone rock absorbs most of the moisture.

The farmers would collect the milk, curdle it with rennet, then scoop the curds by hand into molds. A powder made from grating moldy bread was sprinkled into the curds. At least since the seventeenth century, the mold came from a huge round bread, half wheat and half rye. Probably other breads were used earlier. The bread was stored in the same damp caves that aged the cheese, and in a few weeks it turned blue and was ground to dust for cheese making. The crumbs fermented in the cheese, creating bubbles, which after weeks also started to turn blue.

In 1411, the French Crown granted a patent declaring that only the cheese of Roquefort-sur-Soulzon could be called Roquefort cheese. Roquefort-sur-Soulzon was a tiny village with a few families, located by a rock mass called the Combalou Plateau. In the caves that run under the village, heat and humidity from underground springs are trapped in the rocks. But air constantly circulates from faults in the rock, creating air shaft-like tunnels called *fleurines*. The cheese cellars were built 100 feet into the rock in natural caves moistened by the springs and aired by the fleurines.

The environment of the cellar is cool, extremely humid, and moldy. The temperature is constant, about forty-five degrees Fahrenheit day and night all year. The rock walls, the old hand-hewn wooden beams, the wooden shelving where the cheeses are aged—all are continually slippery wet from the moisture. The rocks offer a kaleidoscope of mold and lichen patterns, and it has been discovered in modern times that this growth is essential to developing the flavor of the cheese.

Again, there is the legend of a founding accident: An absent-minded shepherd boy left his lunch of cheese curd and rye bread in a cave and weeks later discovered Roquefort cheese. Even if the legend is apocryphal, it offers a fair description of how the cheese is made. But for the cheese to be of commercial value, it needs to last for some time. The salt of Aigues-Mortes is rubbed on the top of the cheeses at the outset of aging. Twenty-four hours later, the cheese is turned and the process repeated. The salt melts and begins to work into the cheese. Like the cheese of Parma, Roquefort becomes overly salty, and this unfairly gives the cheese a reputation for saltiness. Alexandre-Balthazar-Laurent Grimod de La Reynière, the eighteenth-century Frenchman said to be the first food journalist, claimed that cheese was a salty snack for drinking. "For those who need to provoke thirst Roque-fort cheese deserves more than any other the epithet of the drunkard's biscuit."

~

THE BASQUES LEARNED how to make hams in their long war with the Celts and then learned to market them in their long peace with the ham-loving Romans. *Jambon de Bayonne*, Bayonne ham, was never made in Bayonne but was shipped from the Basque port of Bayonne at the mouth of the Adour River. It has never been clear, however, if the ham is Basque, though the Basques surprise no one by insisting that it is. Modern France has defined the famous jambon de Bayonne, which was first written about in the sixth century, as a product made in the watershed of the Adour, an area including all of French Basqueland and bits of the neighboring regions of Landes, Béarn, and Bigorre.

One thing is clear. The salt with which it is made is not Basque. It is from Béarn, from a village a few miles from Basque country called Salies-de-Béarn, Béarn saltworks. According to medieval legend, a hunter wounded a wild boar and chased it into the marsh. By the time he found the animal lying in water, it was already preserved in salt. The brine springs of Lüneburg near Hamburg have an almost identical legend, and an ancient ham, supposedly made from the porcine discoverer, is on display in the Lüneburg town hall.

Throughout the watershed of the Adour, shards of pots used in salt making have been found, and some have been dated as early as 1500 B.C. Broken Roman pots have been found within walking distance of Salies-de-Béarn.

Whether or not a wounded boar ever fell in that spot in the

center of Salies-de Béarn, the village has since salted millions of pigs. The town grew up around the mouth of a natural brine spring where a large basin was built to catch the escaping brine. The basin was edged in steps to facilitate approaching it with buckets. The earliest mention of this pool is in the twelfth century, and every narrow, winding street in Salies leads to it.

A town official would test the brine concentration by placing an egg in the pool. When the egg floated, salt making could begin. One or two distributions were possible each week. Some would go to the brine basin themselves with a bucket in hand, but most families hired *tiradous* to gather the brine. The large wooden buckets they used, *sameaux,* were an official unit of measure, each holding ninety-two liters (twenty-four gallons) of saltwater. In every distribution, each house was entitled to twenty-six sameaux.

At the toll of the bell, the tiradous would run down the steps into the brine, which they scooped into their sameaux, and run back to their houses with this weighty, twenty-four-gallon load. They each repeated this twenty-six times, as quickly as possible because they were competing with the other families, and the most concentrated brine from the bottom of the basin would be scooped up by the swiftest tiradous. Weaker brine took longer to evaporate, required more firewood, and so was less profitable.

In front of each house was a stone well into which the tiradous would quickly but carefully pour the brine and then run back for more. A canal of hollowed oak trunks ran under the house to the basement salt-making shop where the brine would be boiled.

The families eligible to participate in this community resource were called *part-prenants*. A part-prenant had to be a descendant of one of the original families, though no one knows exactly when these families originated or how many there were. A code was first written in the Béarnaise language on November 11, 1587, when the tradition was already many centuries old. This code defined the group and stated that descendants had part-prenant rights only if they resided within the ramparts of town. If a woman married "a foreigner"—someone from out of town—her children would only be entitled to a half portion, thirteen sameaux, and their future descendants would receive nothing. But a man could marry an out-of-towner, and he and his heirs would receive full portions. In the fourteenth century, there were 200 part-prenant families, but by the time of the French Revolution, Salies-de-Béarn had 800 part-prenant families.

~

ON THE MEDITERRANEAN coast, west of Aigues-Mortes, in Catalan country near the Spanish border, was the fishing village of Collioure. The people of Collioure lived on selling wine and salted fish. They fished anchovies from May to October on small wooden boats that could sail over the rocks of the shallow harbor, powered by a lateen sail, a triangle of canvas gracefully draped from the mast on a cross spar at a sixty-degree angle. The design dated back to the Phoenicians, but in Collioure they called their fishing boats *catalans* and painted them in brilliant primary colors.

In October, as the anchovy season ended, the wine harvest began on the terraced hills above the town. The wine, called Banyuls, has a dark spicy sweetness that is a perfect counterbalance to the salted anchovies. The people of Collioure worked their patch of vineyard, cutting back the vines, preparing for the next year until the leaves and the shoots came, and then it was May, time to let the grapes grow and fish anchovies. Each family had a catalan for fishing and a patch on the hill for growing grapes. The men went to sea, and the women mended the nets and sold the catch in town.

Most of the catch was put up in salt. Originally, the anchovy salters used local sea salt from Laplame, one of several sea salt operations in the natural ponds along the Catalan coast. But in time the saltworks at the mouth of the Rhône dominated.

In the fourteenth century, an epidemic of bubonic plague, whose delirious victims die within days in excruciating pain, swept through the continent, killing 75 million people—as much as half the population of Europe, according to some estimates. But the fishing village of Collioure was not touched, and it was widely believed that the town was immune because of the presence of huge stockpiles of salt for the anchovies.

Since the time of ancient Greece, anchovies have been the most praised salted fish in the Mediterranean, and since the Middle Ages those of Collioure have been regarded as the best salted anchovies in the world. They are smaller, leaner, more flavorful than their Atlantic cousins. In the Middle Ages, Collioure was also famous for its salted tuna and sardines. The salters were men because their strength was needed to heft the salt. Anchovy filleting was done by women because it required small fingers to rip the tiny fillets off the bones. Freshly caught anchovies were mixed in sea salt and kept for a month. Then the heads were removed and the fish cleaned with no tool other than the swift nimble fingers of the women, who then carefully arranged the fish in barrels with alternating layers of fish and salt. There they remained with a heavy weight on top for about three months. The length of time depended on the size of the fish and the weather, especially the temperature. When the anchovies were ripe, the color of the meat around the bones was a deep pink, almost wine colored, and the brine, produced by the salt melting in the juice it extracted from the fish, turned pink. Unscrupulous anchovy makers dyed their brine pink.

ANCHOVIES: These delicate fish
are preserved in barrels with
bay salt, and no other of the
finny tribe has so fine a flavor.
Choose those which look red and mellow, and the bones must
be oily. They should be high flavored, and have a fine smell; but
beware of their being mixed with red paint to improve their col-
or and appearance.—*Mary Eaton,* The Cook and Housekeepers
Complete and Universal Dictionary, *Bungay, England, 1822*

The French Crown attached such importance to the commercial potential of Collioure salted fish that the town was exempted
from any salt tax. This greatly aided the local anchovy business,
but it was the sort of arbitrary exception that was to make the
French salt tax a political disaster.

CHAPTER TEN

The Hapsburg Pickle

I N G E R M A N Y, T H E Romans had found a land of ancient salt
mines. Tacitus wrote in the first century A.D. that the Germanic
tribes believed the gods listened more attentively to prayers if
they were uttered in a salt mine. But many of those mines had
been destroyed or closed down in the warfare that followed the
disintegration of Rome. As in France, the medieval Church re-
opened them. Monasteries were often located on the sites of ancient mines so that the salt could provide revenue.

Under the direction of the Church, salt mining boomed in the
Middle Ages in the Alpine area from Bavaria into Austria. In
Bavaria, Berchtesgaden and adjacent Reichenhall; across the
Austrian border, Hallein, Hallstatt, Ischl, and Aussee were all
mining the same underground bed of salt. The Austrian part became known as the Salzkammergut, the salt mother lode, a region
of salt mines below green, pine-covered mountains and deep blue
lakes. In the winter, the steep pine forests were completely white

with snow, but underground the temperature in the mines remained moderate.

Underground springs provided brine that could be boiled into salt crystals. Plentiful forests offered cheap energy. Reichenhall, which was still in operation in Roman times, was destroyed in the fifth century either by Attila the Hun or possibly by the local supporters of Odoacer, Rome's final conqueror, the Germanic-Italian who in 476 officially ended the Western Roman Empire. A century later, according to some accounts, the saltworks was reconstructed. According to others, it was rebuilt 300 years later by the archbishop of Salzburg.

Next to Reichenhall was a mountain of salt. On the Bavarian side was Berchtesgaden, and across the border, on the opposite side of the same steep wooded mountain known as Dürnberg, was the ancient Celtic salt mining site of Hallein.

A medieval conflict between the archbishops of Salzburg and the Bavarians over control of the salt mines continued for centuries because Dürnberg mountain contained a Salzburger mine entered on the Hallein side and a Bavarian mine entered from Berchtesgaden. Underground, the two mines were supposedly separated by less than a half mile, but the shafts from Hallein wandered under the border and so Salzburger miners theoretically took Bavarian salt.

The first archbishop of Salzburg had resurrected the ancient Celtic mine in the late eighth century and with this salt revenue had built the city of Salzburg, which did not merge with Austria until 1816. Though Salzburg's territory had gold, copper, and silver, it was salt for which Salzburg repeatedly fought. The wealth from salt gave Salzburg its independence.

In the seventeenth century, an archbishop named Wolf Dietrich tried to dominate the salt market by dramatically lowering the selling price for salt from his mines, especially Dürnberg. For a time Dietrich made tremendous profits, some of which were used to build grand baroque buildings in Salzburg. Bavaria retaliated by banning trade with Salzburg, and this eventually led to a "salt war," a conflict which Dietrich lost. This defeat was disastrous for Dürnberg and its village of Hallein because, for a time, they were excluded from much of the regional salt trade. It was even more disastrous for Archbishop Wolf Dietrich, who was removed from his Church post and, after five years in prison, died in 1617.

The relationship between the two sides of Dürnberg mountain was not resolved until after Salzburg became a part of Austria, when, in 1829, a treaty between Bavaria and Austria allowed Austrians to mine salt up to one kilometer beyond their border. In exchange, 40 percent of mine workers had to be Bavarian, and Bavaria could fuel its pans with trees chopped on the Austrian side. Though fuel had been plentiful in the Middle Ages, after centuries of mining, procuring wood had become an important issue.

IN 1268 AND possibly earlier, a new technique was used to mine rock salt. Instead of miners carrying chunks of rock out steep shafts in baskets slung on their backs, and then crushing the rock into salt, water was piped into a dug-out vein of rock salt. The water quickly became a dense brine, which was then piped out of the mountain to the village of Hallein, where it was boiled down into crystals over wood-burning fires.

Eventually, the idea became a more sophisticated system known in the Salzkammergut as *sinkwerken*. A sinkwerk was an underground work area in which the surrounding salt and clay were mixed with water in large wooden tanks. The solution then moved down wooden pipes to iron boiling pans.

Hallein is a village pressed between its two sources of wealth, the rough, rock-faceted Dürnberg Mountain and the Salzach River. The Salzach is a tributary of the Danube, and the brown Danube runs, with its tentacles of tributaries, from west of Bavaria through central Europe to the Black Sea. The salt could be boiled in cylindrical molds, much as it still is in Saharan Africa, and the cylinders could be loaded in barges that traveled the Salzach to Passau, where it entered the Danube, to be traded in Germany or central Europe.

But much of Hallein's salt was for local use, traveling by river only to Passau, where it was packed on wagons to be sold in the region. Transporting on land was expensive because tolls were es-

Diagram of sinkwerken in Dürnberg Mountain shown from the Bavarian side in Berchtesgaden. Deutsches Museum, Munich

tablished along the roadways for wagons carrying salt. The inevitable response was a network of paths over rugged mountain
passes for smugglers carrying illegal salt, which they could sell for
less because they paid no tolls.

RIVERS WERE ESSENTIAL to central European saltworks.
Halle in central Germany and Lüneburg in the north, with its famous founding ham, had the advantage of the Elbe with its
mouth at the North Sea port of Hamburg. In the late fourteenth
century, the Lüneburgers built a canal, the Steckenitz Canal, to
connect their salt to the Elbe. They did this not to move salt to
nearby Hamburg but to ship it to Lübeck, on a tributary to the
Baltic, because Lübeck was the trading center of the Hanseatic
League.

In the Middle Ages, no German salt enjoyed as great an international reputation as that of Lüneburg, and the Hanseatics shipped it to the herring fisheries of southern Sweden, to Riga, to Danzig, and throughout the Baltic. At a time when the Hanseatic League was considered the guarantor of quality, Lüneburg salt was considered the Hanseatic salt. Lesser German saltworks would fraudulently mark their barrels with the word *Lüneburg* to obtain entry to foreign markets.

The salt in Lüneburg, Halle, and other German saltworks was made by drawing the brine in buckets and carrying it to boiling sheds, where it was dumped in a huge rectangular iron pan. The pan sat on a wood-burning furnace. Blood was added, which caused a scum to rise with boiling. The scum drew impurities and was skimmed off with care. The salt maker needed to continually stir the liquid. Shortly before crystallization, beer was added to further draw impurities from the crystals, which were then placed to dry in conical baskets.

With the pans in use twenty-four hours a day, except for a once-a-week cleaning, the entire operation required only three people: a master salter, an assistant, and a boy to stoke the furnace. This staff was often simply a man, his wife, and a son. It was easy for a family to go into the salt business. But in Lüneburg the business did not remain in the family because, one by one, Hanseatic merchants bought them out and gained control of a single large saltworks.

THE SALZKAMMERGUT DEVELOPED its own salt-mining culture. Saint Barbara was its patron saint, and miners observed her day, December 4, by performing traditional dances in their own dress uniforms, which, by the nineteenth century, included a black wool jacket with brass buttons and epaulets and a black velvet hat with silk buttons and a gold emblem of two crossed pickaxes.

The Dürnberg mine has nineteen miles of tunneling. The main tunnel was built in 1450, but the current timber shoring is only 100 years old. The tunnels were built with seven-foot ceilings and wide enough for a man to walk through comfortably. But the pressure of the mountain's weight slowly compresses them. It was in such compressed ancient shafts that medieval miners discovered the remains of their Celtic predecessors. Today, one of the 400-year-old tunnels is only about eighteen inches wide in parts. Another seventeenth-century tunnel is about three feet wide.

The tunnels, shored up with timber, have walls of dark rock with white streaks of salt crystals. In some spots the rock is spotted with fossils of shellfish and other marine life. Miners would ride on steep, smooth wooden slides, which propelled them at considerable speed down to tunnels sometimes as far as 100 feet

Engraving of miners descending a shaft in the Dürnberg salt mine on a slide. The rope on their right is used as a brake. Salinen Austria, Dürnberg

lower. Some of the slides are more than 350 feet long. A cable on the right side provided a brake for the practiced gloved hands of the miners.

Dürnberg has been hosting visitors since at least the end of the seventeenth century, when tours were a special treat for elite guests of the archbishop of Salzburg. Centuries ago it was realized that the slides could be fun. A miner at the top and five or six guests, all hugging each other, slid down as if they were in a roller coaster car. The mine also has about twenty-five underground lakes for boat rides.

MUCH OF THE salt of central Europe eventually came under the control of the Hapsburgs. From their beginnings in tenth-century Alsace, and as their rule spread across central Europe, the Hapsburg family controlled salt mines. In 1273, a Hapsburg became Rudolf I, king of Germany, who enlarged his holdings by conquering Bohemia. The Hapsburgs gained control of the Danube, Silesia, Hungary, and the southern region of Poland known as Galicia. For a time they even had Spain and all its New World possessions as well as the Netherlands, Naples, Sardinia, Sicily, and Venice.

The Hapsburgs established a salt monopoly, controlling production, transport, and wholesale trade. Bohemia, one of the wealthier regions of Europe, was saltless, an eager market for other Hapsburg holdings in what is today Germany, Austria, and southern Poland.

Hungary was another salt-poor region that came under Hapsburg rule. In sixteenth-century Hungary, with an economy based on the export of food, there were only four important food imports: spices, wine, herring, and salt. Much of the export of food depended on the import of salt. Pig fat was a staple for both eating and preserving other food. From the seventeenth century on, fat was included in wages. A high-fat diet was considered a sign of wealth, and city people luxuriated in more fat than peasants. An 1884 study showed that rural Hungarians ate an average per capita of forty pounds of cured—salted or smoked—fat, whereas city dwellers consumed an average per person of fifty-six pounds of fat. This does not include the significant amount of rendered animal fat that was eaten like butter, not to mention butter itself.

Cooking with melted fat rather than preserved pieces was an eighteenth-century innovation—a refinement for the upper classes. The traditional fat was made by opening a freshly slaughtered pig and removing whole the thick outer layer of fat. This was then preserved in dry salt, after which it was smoked, except in the Great Plain, the flat grain-growing region east of the Danube, where it was air-dried. Peasants made thick soups that began by melting down pieces of this salted fat, which produced an oil for frying the rest of the soup's ingredients, and cracklings which were sprinkled on top.

⌒

SOUTHERN POLAND WAS the site of ancient springs where as early as 3500 B.C., brine was gathered and boiled in clay pots. But gradually these springs dried up. In 1247, miners began digging in the earth to get at the rock salt that had hardened at the sources of the brine. In 1278, the Polish Crown took possession of the mine but leased its operation to a succession of entrepreneurs, which included Poles, both Jewish and Christian, French, Germans, and Italians. They made payments to the controlling monarch and offered salt at discount rates to aristocracy.

At first, salt miners, often prisoners of war, were worked to death in slave conditions. Not until the fourteenth century, when free men began working the mines, did it become less than a death sentence. In the sixteenth century, the mines went deeper, and huge pulley systems powered by teams of eight horses hoisted the salt to the surface. Horses that were brought in to work the mines spent their entire lives below ground.

There are mountains in which the salt goes down very deep, particularly at Wieliczka and Bochnia. Here on the fifth of January, 1528, I climbed down fifty ladders in order to see for myself and there in the depths observed workers, naked because of the heat, using iron tools to dig out a most valuable hoard of salt from these inexhaustible mines, as if it had been gold and silver.—*Olaus Magnus,* A Description of the Northern Peoples, *1555*

The Polish Crown earned one-third of its annual revenues from the salt of these two mines near Cracow, Wieliczka and Bochnia.

In 1689, the mines began offering miners daily Catholic services at their underground place-of-work. The miners of Wieliczka began carving religious figures out of rock salt. Three hundred feet below the surface, miners carved a chapel out of rock salt with statues and bas-relief scenes along the floor, walls, and ceiling. They even fashioned elaborate chandeliers from salt crystals.

Increasingly, the mine had visitors. In the early seventeenth century, as in Dürnberg, the Crown began to bring special guests, mostly royalty. They came to dance in ballrooms, dine in carved dining rooms, be rowed in underwater lagoons. In 1830, the Wieliczka Salt Mine Band, which still performs, was started because of the quality of the acoustics in the mine.

The Wieliczka mine and that of nearby Bochnia were near the Vistula, which flowed a few miles north to Cracow and then on to Warsaw and finally to the Baltic. Any salt with a water route to the Baltic had a huge market. But the Baltic port also meant that the coarse, dark gray rock salt of southern Poland had to compete with sea salt from France and Portugal. The Portuguese sold their Setúbal salt to the Hanseatics, who sold it in Holland and Denmark. By the sixteenth century, cheap, white Setúbal salt had also become popular in Poland and other Baltic countries. The Polish Crown responded by protecting its own salt with a ban on the import of all foreign sea salt.

Entertaining visitors in the Grand Hall in the Wieliczka salt mine in 1867. The walls, ceiling, floor, chandeliers, and statues are all made from salt. Culver Pictures

﹏

IN 1772, POLAND was partitioned between Austria, Prussia, and Russia—vanished as a nation until after World War I. In acquiring the Galicia region, the Austrian Hapsburgs gained control of Wieliczka and Bochnia. The salt of these mines was sold not only in Poland but throughout the Hapsburg Empire and in Russia. The huge nation of Russia had a considerable demand for salt, especially to preserve meat and vegetables through a long and barren winter. Salting or corning beef in most societies was reserved for lesser cuts such as the brisket, which is the breast cut under the first five ribs, or the round, the toughest leg cut. But in Russia, beef was often frozen solid in the ground and sawed up with little regard for different cuts.

Salt being transported by camel-drawn carts to the railroad at Lake Baskuntschak, in the southern Urals of Russia, circa 1929. Culver Pictures

The following recipe comes from *A Gift to Young Housewives,* by Elena Molokhovets. Molokhovets and her book, which she continually revised between 1861 and 1917—poignant years wedged between the emancipation of the serfs and the Communist revolution—were well known in Russian households

SOLONINA (SALTED BEEF)

Use a towel to rub off any blood from freshly slaughtered beef. This must be done while the carcass is still warm because the blood very quickly spoils the meat. Remove the very large bones, weight the meat, and rub it all over with salt that has been dried in the oven and mixed with saltpeter and spices. Lay out the meat on a table to cool completely. Then pack into small barrels, placing the large pieces in the middle and small half-pound pieces around the edges so as not to leave any gaps. Press the meat lightly with a pounder. Sprinkle salt, saltpeter, bay leaves, rosemary, and allspice on the bottom of the barrel and over each layer of meat as the barrel is filled. When the barrel is full, cover it with a lid, seal with tar on all sides, and keep in a [warm] room for two to three days, every day turning the barrel over, first on one end then on the other. Transfer the barrel to the cold cellar, and then turn it over twice a week. After three weeks, store the barrel on ice.

Use the following proportions of salt and spices. For one and a half poods of meat (1 pood = 36.113 U.S. pounds so this is about 54 pounds), use two and a half pounds well-dried salt, six zolotniki (1 zolotnik = about 1 U.S. teaspoon) saltpeter, and three lots (1 lot = about one half ounce or a tablespoon) each coriander, marjoram, basil, bay leaf, allspice, and black pepper. Add garlic if desired. Sprinkle a little extra salt into those barrels that will be used later.

The barrels must be small and made of oak, because when a barrel is unsealed and the meat is exposed to the air, it soon spoils. The barrels must be sealed all over, to prevent the juice from leaking.

Before salting the meat, the barrels should be soaked and disinfected.—*Elena Molokhovets,* A Gift to Young Housewives

THE MOST COMMON salt-cured vegetables from Alsace to the Urals were cucumbers and cabbage—pickles and sauerkraut. The importance in central Europe of lactic fermentation of vegetables, commonly known as pickling, is best expressed by the Lithuanians, who recognize a guardian spirit of pickling named Roguszys.

In any pickling it is crucial to prevent exposure to the air, which leads to rot rather than fermentation. This is accomplished either by careful sealing, as in the beef recipe above, or by keeping the food submerged in brine by weights. Sand is used as the weight in the following recipe.

SOLENYE OGURTSY (SALTED CUCUMBERS)

Dry out very clean river sand and pass it through a fine sieve. Spread a layer of this sand, the thickness of your palm, on the bottom of a barrel. Add a layer of clean black currant leaves, dill, and horseradish cut into pieces, followed by a layer of cucumbers. Cover the cucumbers with another layer of leaves, dill, and horseradish, topped with a layer of sand. Continue in this manner until the barrel is full. The last layer over the cucumbers must be currant leaves, with sand on the very top. Prepare the brine as follows: For one pail of water, use one and a half pounds of salt. Bring to a boil, cool, and cover the cucumbers completely with the brine. Replenish the brine as it evaporates. Before any kind of salting, cucumbers must be soaked for 12–15 hours in ice water.—*Elena Molokhovets,* A Gift to Young Housewives

Copper ions could leach into the food from copper pans, brightening colors, especially the green of vegetables. It made pickles look beautiful but troubled the digestion, which has little tolerance for copper. Molokhovets gave this warning:

> Purchased cucumbers are sometimes very attractive, that is, green as a result of being prepared in an untinned copper vessel, which is extremely harmful to your health. To check whether the greenness of the cucumbers is really a result of this preparation, stick a clean steel needle into a cucumber. The needle will turn a copper color in a short time if the cucumbers have been adulterated.

The amount of salt used in sauerkraut in Russia and Poland depended on the economic status of the family. Families that could afford to do so used not only salt but seasoning, such as caraway seeds, dill, and in southern Poland, cherry leaves. In Moravia apples and onions were added. The Moravians also added bread to speed up fermentation. In Poland, making sauerkraut was a community ritual every fall after the potato harvest. Women would slice the cabbage, scald it in hot water, and place it in barrels—sometimes in wood-lined ditches in the ground. Then men would pound it with clubs or by stamping their feet to prevent air bubbles, which could cause rot. Women then covered the cabbage with linen and lids weighted by heavy stones to make sure the vegetable remained completely submerged. An annual dance marked the occasion when the year's supply of sauerkraut had been covered. But the work was not finished. The cloth had to be periodically cleaned, mold scraped off the lids, and water added to keep the cabbage submerged for two weeks before it could be stored in a cellar for the entire winter.

In Poland and Russia, sauerkraut was an ingredient to be used in other dishes. Whole cabbages would be included with the sliced ones because the whole pickled leaves were needed for *go-labki*, which means "*pigeon*" but is actually cabbage stuffed with buckwheat and meat. The brine was used as a soup base. Sometimes the sauerkraut was squeezed for the juice and the cabbage pieces discarded.

The Polish national dish, *bigos*, is sauerkraut to which meat, bacon, pickled plums, and other fruits are added. The dish, a kind of Polish choucroute, was made in past centuries in a clearing in the forest. Hunters, generally aristocrats, would come to the clearing to add their game. *Pan Tadeusz*, a poem of rural life in Lithuania, today considered the Polish national poem, describes bigos.

> *The bigos is being cooked. No words can tell*
> *The wonder of its color, taste, and smell.*
> *Mere words and rhymes are jingling sounds, whose sense*
> *No city stomach really comprehends.*
> *For Lithuanian food and song, you ought*
> *To have good health and country life and sport.*
>
> *But bigos e'en without such sauce is good,*
> *of vegetables curiously brewed.*
> *The basis of it is sliced sauerkraut,*
> *Which, as they say, just walks into the mouth;*

Enclosed within a caldron, its moist breast
Lies on the choicest meat, in slices pressed.
There it is parboiled till the heat draws out
The living juices from the cauldron's spout,
and all the air is fragrant with the smell.
—Adam Mickiewicz, *Pan Tadeusz*, 1832

CHAPTER ELEVEN

The Leaving of Liverpool

IN THE LIST of great rivers that played essential roles in the history of salt—the Yangtze, the Nile, the Tiber and the Po, the Elbe and the Danube, the Rhône and the Loire—a gurgling mud-bottomed waterway that flows for only seventy miles from the English midlands to the Irish Sea has to be included: the River Mersey.

The importance of the Mersey lay not in the goods it carried those few dozen miles into England but in what it carried from England to the world. The last three miles of the river form a sheltered, deepwater harbor, and in 1207, King John granted permission for a town to be built there, which was called Liverpool. Originally Liverpool was the port that connected Ireland with England. But in time it became England's most important port after London. It was the port of West Indian sugar, the port of the slave trade, the Industrial Revolution port that brought iron to coal and then shipped out steel. But before any of this, it was the port of English salt, Cheshire salt, or, as it became known all over the world, Liverpool salt.

~

WHEN THE ROMANS came to England in A.D. 43, they found the Britons making salt by pouring brine on hot charcoal and scraping off the crystals that formed. To the Romans, this was a sign of pitiful backwardness, and being the model imperialists, they taught these primitive locals the right way to make salt—by evaporating brine in earthen pots and then smashing the pots to expose white cakes of salt. The Romans started saltworks along the entire east coast. They established London in their first year in Britain, and, remembering how Ostia provided for the growth of Rome, they developed saltworks in Essex to provide for what they hoped would become a major port city on the Thames.

The Romans were drawn to the thick forest of northwestern England, probably for fuel, because the peat they had been using to evaporate brine on the coast was becoming scarce. In the northwest they found a place the locals knew by the Celtic name Hellath du, which meant "black pit." By the time the Romans reached this area, later known as Cheshire, it had been producing salt for centuries. The earliest evidence of salt making in Cheshire, pottery fragments dated to 600 B.C., shows that the Britons had long known the "new" Roman technique.

The neighboring area, what is today North Wales, had silver mines. When the silver was extracted, lead remained, which the Romans used to make huge pans, some weighing more than 300 pounds, for boiling brine in Hellath du, the first pan-evaporated salt in England. The locals too learned to evaporate in lead pans, but preferred a nearby location called Hellath Wenn, white pit, and not by coincidence this produced a whiter salt.

In time, Hellath du acquired the Anglo Saxon name North-wich, northern saltworks. Anglo Saxons called a saltworks a *wich*, and any place in England where the name ends in "wich" at one time produced salt. Hellath Wenn became Nantwich, and be-tween Nantwich and Northwich was Middlewich.

By the ninth century, the area by the mouth of the Mersey, Cheshire, had become an important salt-producing region. The commercial center was Chester, where, in the eleventh century, the Roman-built fort was the last Saxon fortress to fall to William the Conqueror, completing the Norman conquest of England. In 1070, to crush the resistance, the Normans destroyed Chester and its saltworks, and in the decades that it took Chester to re-build, Droitwich, south of Cheshire in Worcestershire, emerged as England's leading salt producer.

CHESTER WAS ON the River Dee, which had an estuary that provided a deepwater port similar to that of the Mersey. Once Liverpool was founded on the Mersey, the two towns, with their two parallel rivers only a few miles apart, were competitors until the Dee began silting up and all the trade shifted to Liver-pool.

For centuries, Bristol was a more important port, even a more important salt port, than Liverpool. This was due not to the ex-portation of British salt, but to the many ships carrying imported Portuguese and French sea salt that docked there. British salt-works could not provide the sea salt needed for British fisheries. Even when the English made a special high-quality salt for the cure of the best herring, a salt called white on white, they made it with French sea salt, dissolving the French salt in water and reevaporating it to remove impurities.

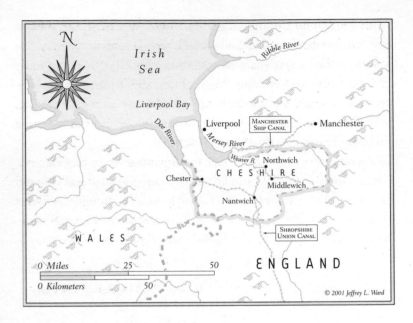

The market for salt fish proved more durable than the religious convictions that created it. Even after 1533, when Henry VIII broke with the Roman Church, a lenten meat eater was still subject to an array of penalties including three months' imprisonment and public humiliation. By this time the motivation was less religious than economic—the government wanted to support the fishing industry. A 1563 proposal to extend the lean days to twice a week, adding Wednesday to Friday, was supported by the argument that it would build up the fishing fleet. It took twenty-two years of debate, but the idea of a second fast day was finally dropped in 1585. The English people were growing weary of the fast laws, and the Church adapted. The selling of permits to eat meat on fast days was becoming a profitable source of Church revenue.

In 1682, John Collins, an accountant to the British Royal Fishery, wrote a book called *Salt and Fishery, Discourse Thereof,* inspired by his seven years at sea, from 1642 to 1649, primarily serving with the Venetian fleet fighting the Turks. During this time, he was obliged to eat badly salted meat, evidently rotting, which he said "stunk." This experience, he said, "begat in me a curiosity to pry into the nature of salt."

Among his many recipes was the following for curing salmon. The recipe would still be good today, assuming a fifteen-year-old boy were available for long periods of jumping. Though the fish is from the Scottish-Northumberland border, the salt specified, as was usually the case for curing fish, was French sea salt:

The salmon cured at Berwick. As described by Benjamin Watson, merchant.

1. They are commonly caught from *Ladiday* [March 25, Feast of the Annunciation when the angel came to Mary] or *Michael-mas* [September 29, Saint Michael's Day] either in the river Tweed or within three miles or less off at sea against Berwick.

2. Those caught in the upper part of the river. Brought by horseback to lower part. And those on the lower part thereof on boats to Berwick, fresh.

3. Then they are laid in a pav'd yard, where for curing there are ready 2 splitters and 4 washers.

4. The splitters immediately split them beginning at the tail and continuing to the head, close by the back fin, leaving the Chine of salmon on the under side [the belly intact], taking the guts clear out and the gils out of the head, without defacing the least fin and also take out a small bone from the underside, whereby they get to the blood to wash it away.

5. Afterward the fish is put into a great tub, and washed outside and inside and scraped with a mussuel shell or a thin iron like it; and from thence put into another tub of clean water, where they are washed and scraped again, and from thence taken out, and laid upon wooden forms, there to lie and dry for four hours.

6. Thence they are carried into the cellars, where they are opened, or layed into a great vat or pipe with the skinside downward and covered all over with French salt and the like upon another lay and so up to the top and are there to remain six weeks. In which time tis found by experience, they will be suffeciently salted.

7. Then a dried calves' skin is to be laid on at the top of the Cask, with Stones upon it to keep them down; upon the removal thereof, after 40 days or thereabouts, there will appear a scum at the top about two inches deep, to be scum'd off or taken away.

8. Then the fish is to be taken out and washed in the pickle, which being done, they are to be carefully laid into barrels, and betwixt every lay, so much salt sprinkled of the remaining melted salt in the vats, as will keep them from sticking together. And after the barrel is one quarter full, is to be stamped or leaped upon by a youth of about 15 years old or thereabouts, being coverede with a calves skin, the like at half full, and also when quite full.

9. Then a little salt is to be laid on the top and so to be headed up; and then the Cask is to be hooped by the cooper and blown til it be tight.

10. Then a bunghole to be made in the middle of the barrel, about which is to be put a ruff or roll of clay, to serve as a Tonnel whereby frequently to fill the barrel with the pickle that is left in the vat, which will cause the oyle to swim; which

ought to be frequently scummed off, and serves for greasing of wool. And thus after 10 or 12 days to be bounded up as sufficiently cured, and fit for exportation.—*John Collins,* Salt and Fishery, Discourse Thereof, *1682*

EVEN WITHOUT FISH, Cheshire salt had ample uses. Crops to feed both humans and livestock could only be provided until the November harvest. The animals would then be slaughtered and salted to last until spring grasses could support a new herd. Animals were slaughtered on Martinmas, November 10, the Saint Day of Martin, an austere Roman soldier in Gaul who converted to Christianity and became the patron saint of reformed drunkards. Pre-Christian religions also marked November 10 as the day on which animals were slaughtered and salted for the winter, followed by a celebration for which, if they too converted, Saint Martin could grant forgiveness.

English food was extremely salted. Bacon had to be soaked before using.

Take the whitest and youngest bacon and cutting away the sward [rind] cut the collops [slabs] into thin slices, lay them in a dish, and put hot water into them, and so let them stand an hour or two, for that will take away the extreme saltinesse.—*Gervase Markham,* The English Huswife, *1648*

Vegetables were also put up in salt to be used throughout the winter, and they too had to be refreshed before use. John Evelyn, a notable seventeenth-century English scholar who argued for more vegetables and less meat, gave this recipe for preserving green beans:

Take such as are fresh young and approaching their full growth. Put them into a strong brine of white-wine vinegar and salt able to bear an egg. Cover them very close, and so will they be preserved twelve months: but a month before you use them, take out what quantity you think sufficient for your spending a quarter of a year (for so long the second pickle will keep them sound) and boil them in a skillet of fresh water, till they begin to look green, as they soon will do. Then placing them one by one, (to drain upon a clean course napkin) range them row by row in a jarr, and cover them with vinegar, and what spices you please; some weight being laid upon them to keep them under the pickle. Thus you may preserve French beans, harico's etc. the whole year about.—*John Evelyn,* Acetaria: A Discourse of Sallets, *1699*

Butter was also very salty. A 1305 recipe from the estate of the bishop of Winchester called for a pound of salt to be added for every ten pounds of butter. This would produce a butter as salty as Roman garum. The salt was to preserve the butter rather than for taste, and numerous medieval writers gave recipes for desalting butter before using, which often entailed mixing with fresh butter.

Butter has the same improbable myth of origin as cheese, that it accidentally got churned in the animal skins of central Asian nomads. Easily spoiled in sunlight, it was a northern food. The Celts and the Vikings, and their descendants, the Normans, are credited with popularizing butter in northern Europe. Southerners remained suspicious and for centuries maintained that the reason more cases of leprosy were found in the north was that northerners ate butter. Health-conscious southern clergy and noblemen, when they had to travel to northern Europe, would guard against the dreaded disease by bringing their own olive oil with them.

With no refrigeration, unsalted butter quickly becomes rancid. Even the butter sold as "sweet" was lightly salted. The English did have a specialty called May butter, which was fresh spring butter left unsalted in the sunlight for days. The sunlight would destroy the carotene, turning the butter white, and along with the pigment would go all of its vitamin A. It would become rancid and, no doubt, smell rancid. But inexplicably, in the Middle Ages May butter was considered a health food.

In the Middle Ages, yellow flowers of various species were salted and kept in earthen pots and beaten to extract a juice to color butter that had lost its carotene. Later, after Columbus's voyages, annatto seeds were used. These seeds are still used by large American dairies, not to conceal rancid butter but because they believe the consumer wants a consistent dark yellow color.

The English passed laws against selling rancid butter. A 1396 law outlawed the use of salted yellow flowers. In 1662, a butter law was passed in England to establish standards. It allowed mixing rancid butter with good and specified that butter could only be salted with fine, not coarse-grained, salt, and it had to be packed with the producers' first and last name clearly marked.

To preserve butter fresh for long keeping.

Make a brine as before described (salt enough to float an egg) and keep the butter sunk in it. About the beginning of May I caused this to be put into practice and potted up many lumps of butter bought fresh out of the market, and they all kept sweet, fresh.—*John Collins,* Salt and Fishery, Discourse Thereof, *1682*

The Church did not allow butter to be eaten on fast days because it came from cows. But it also earned enormous profits selling special dispensations to affluent people who could not bear going without butter for the forty days of Lent. Lent aside, butter was cheap food and was more popular with the poor than the rich. Because of the heavy salting, it was available most of the year. Beginning in the sixteenth century, it was even included in the rations of the Royal Navy.

Determined to make butter more than a luxury for a rural elite, northern Europeans consistently tried to preserve it in salt. But getting good, properly preserved butter remained a problem until refrigeration was invented. In fact, the first experiments in refrigeration were not with fish or meat but with everyone's favorite luxury—butter.

Cheese, the more successful way to preserve milk and cream, was also a popular salted food of the poor, though only the wealthy sampled the full array of English cheeses—some 150 varieties (or at least this many were remaining in the 1970s when British cheese enthusiast Patrick Rance went on a crusade to save traditional English-cheese making).

Cheshire, not surprisingly for a place with both dairy herds and saltworks, produced a great deal of cheese. Cheshire is the oldest known variety of English cheese and is thought to be more representative of a medieval English cheese than is cheddar or the blue-veined Stilton. A hard cheese, though not as hard as cheddar, it has a distinct flavor thought to come from the salty earth grazed upon by Cheshire cows.

By the seventeenth century, the English had discovered that salted anchovies would melt into a sauce. This practice may have existed centuries earlier on the continent, but in the seventeenth and eighteenth centuries, anchovy sauces became extremely popular. Grimod de La Reynière, a great eighteenth-century anchovy sauce enthusiast, wrote, "When this sauce has been made well, it would make you eat an elephant."

In 1668, a French writer, Pierre Gontier, stated that "anchovies are put in salt in order that they may be preserved, and they become garum." Certainly, the English at the time used anchovy sauce very much like garum—a liquid of preserved fish that was added to meat and other dishes as a salty seasoning.

In eighteenth-century England, anchovy sauce became known as ketchup, katchup, or catsup.

To make English Katchup

Take a wide mouth'd bottle, put therin a pint of the best white wine vinegar, putting in ten or twelve cloves of eschalot peeled and just bruised; then take a quarter of a pint of the best langoon white wine, boil it a little, and put to it twelve or fourteen [salt cured] anchovies washed and shred, and dissolve them in the wine, and when cold, put them in the bottle; then take a quarter of a pint more of white wine, and put it in mace, ginger sliced, a few cloves, a spoonful of whole pepper just bruised, and let them boil all a little; when near cold, slice in almost a whole nutmeg, and some lemon peel, and likewise put in two or three spoonfuls of horse radish; then stop it close, and for a week shake it once or twice a day; then use it; it is good to put into fish sauce, or any savory dish of meat; you may add it to clear liquor that comes from mushrooms.—*Eliza Smith*, The Compleat Housewife, *posthumous 16th edition, 1758*

Ketchup derives its name from the Indonesian fish and soy sauce *kecap ikan*. The names of several other Indonesian sauces also include the word *kecap*, pronounced KETCHUP, which means a base of dark, thick soy sauce. Why would English garum have an Indonesian name? Because the English, starting with the medieval spice trade, looked to Asia for seasoning. Many English condiments, even Worcestershire sauce, invented in the 1840s, are based on Asian ideas.

Whether it is called garum, anchovy sauce, or ketchup, a large dose of salt was an essential ingredient. Margaret Dods cautioned in her 1829 London cookbook that "catsups, to make them keep well, require a great deal (of salt)." The salt in ketchup originally came from salt-cured fish, and most early anchovy ketchup recipes, such as Eliza Smith's, do not even list salt as an ingredient because it is part of the anchovies. But the English and Americans began to move away from having fish in their ketchup. It became a mushroom sauce, a walnut sauce, or even a salted lemon sauce. These ketchups originally included salt anchovies, but as Anglo-Saxon cooking lost its boldness, cooks began to see the presence of fish as a strong flavor limiting the usefulness of the condiment. Roman cooks would have been appalled by the lack of temerity, but Margaret Dods adds at the end of her walnut ketchup recipe:

> Anchovies, garlic, cayenne, etc. are sometimes put to this cat-sup; but we think this is a bad method, as these flavours may render it unsuitable for some dishes, and they can be added ex-tempore when required.—*Margaret Dods,* Cook and Housewife's Manual, *London, 1829.*

Ketchup became a tomato sauce, originally called "tomato ketchup" in America, which is appropriate since the tomato is an American plant, brought to Europe by Hernán Cortés, embraced in the Mediterranean, and regarded with great suspicion in the North. The first known recipe for "tomato ketchup" was by a New

Jersey resident. All that is certain about the date is that it had to be before 1782, the year his unfashionable loyalty to the British Crown forced him to flee to Nova Scotia.

The first published recipe for tomato ketchup appeared in 1812, written by a prominent Philadelphia physician and horticulturist, James Mease. Already in 1804 he had observed, employing the term used for tomatoes in the United States at the time, that "love apples" make "a fine catsup." Mease said that the condiment was frequently used by the French. The French have never been known for their fondness for tomato ketchup, so it is thought, given the date, that the French he was referring to, were planter refugees from the Haitian revolution. To this day, a tomato sauce is commonly used in Haiti and referred to as *sauce creole*.

LOVE-APPLE CATSUP

Slice the apples thin, and over every layer sprinkle a little salt; cover them, and let them lie twenty-four hours; then beat them well, and simmer them half an hour in a bell-metal kettle; add mace and allspice. When cold, add two cloves of raw shallots cut small, and half a gill of brandy to each bottle, which must be corked tight, and kept in a cool place.—*James Mease*, Archives of Useful Knowledge, *Philadelphia, 1812*

Ketchup remained a salted product. Lydia Maria Child, in her 1829 Boston cookbook, *The American Frugal Housewife,* advised in making tomato ketchup, "A good deal of salt and spice is needed to keep the product well."

⌒

AT THE END of the seventeenth century, Cheshire salt was still produced from two brine pits in Middlewich, one in Nantwich, and one at Northwich. If a Chinese salt producer had gone to Cheshire in the 1500s, he would have been appalled by the primitiveness of the technology. Shirtless men climbed down ladders into the pits, filled leather buckets with brine, and climbed out to dump the brine in wooden troughs. Then a web of pipes and gutters channeled the brine to the many salt makers in the area. But by 1636, an account of a visit to the wiches mentioned that pumps had just been installed in Nantwich to raise the brine.

In the eighteenth century, life in England began to change. England experienced an extremely favorable shift in climate that allowed longer growing seasons and cheaper food. With food prices lowered, many English farms failed. Failed farms in turn created a workforce for industry.

The English, before anyone else, believed industry was the answer to all problems. Agro-industry, which abandoned the goal of producing the best food and strived to produce the most per acre, was an English invention. Wheat crops increased enormously. New feed, such as turnips, kept livestock eating all year. Starting with Jethro Tull's seed-planting drill of 1701, which planted three rows at once, a new agricultural invention, a new crossbred plant, a new strain of livestock, or a new tool was invented almost every year in eighteenth-century Britain. This was the beginning of modern agriculture, a system that would produce enormous surpluses of food in industrialized nations and still fail to end hunger in the world.

These new developments in agriculture meant that food could be produced throughout the year, which meant less dependency on salt. Less salt seemed a modern idea. But salt production was increasing. Just as the Roman occupiers had run out of peat, English saltworks were now running out of trees.

The fule which was heretofore used was all wood, which since the iron works is destroyed, that all the wood at any reasonable distance will not supply the works with one quarter of the year; so that now we use almost all pit coal which is brought to us by land, from 13 or 14 miles difference.—*Dr. Thomas Rastel*, Droitwich, *1678*

By 1650, little was left of the forests of Cheshire. The lead pans by this time were each almost as big as a room and were installed on top of coal-fired furnaces. Hauling coal to Cheshire became a major expense of salt production. Salt makers began to wonder if there might not be coal underneath Cheshire. They were surrounded by coal regions. In Whitehaven, not far north of them in Cumberland, and farther up near Glasgow at the mouth of the Clyde, salt was made and sold at a much lower price than Cheshire's product because these saltworks had their own coal fields.

Elizabeth I, concerned about England's dependence on French salt, had guaranteed state-controlled markets to salt producers along the Tyne in Northumberland. She had chosen that region for stimulating production because it had coal for cheap fuel.

Cheshire had salt, a river, and an Atlantic port. Its salt makers could have provided salt to a world in which British influence was rapidly expanding—if only they had cheap fuel. The Cheshire salt producers went coal prospecting. In 1670, John Jackson prospected for coal on the estate of William Marbury near Northwich and, at a depth of only 105 feet, found a bed of solid rock salt and no coal at all.

The Royal Society published the news first with great excitement. Had Jackson discovered the source of underground brine? Was it a buried seabed? In 1682, John Collins wrote of Cheshire, "These springs being remote from the sea are conceived to arise from rocks or Mines of salt under the earth, the which are moistened by some channels or secret passages under ground."

But Marbury, disappointed that Jackson had found no coal, did not think to mine the rock salt and went bankrupt in 1690. In 1693, another Cheshire landowner, Sir Thomas Warburton, found rock salt under his estate, and four years later he owned one of four rock salt mines that opened in Cheshire.

Rock salt did not need fuel, but the immediate reaction of the Cheshire brine boilers was to lobby parliament for a bill banning the mining of rock salt. They believed that the discovery would change the nature of Cheshire, that the small-scale entrepreneur with a modest investment in a well and some lead pans would be pushed out by large and well-capitalized mining companies.

But with the discovery of rock salt, the growing salt industry gained the economic importance to persuade the government to construct canals. Between 1713 and 1741, the government built a network of waterways linking the saltworks with the Mersey. By the end of the century, salt refineries were being established along the Mersey, and a salt warehouse was built on the Liverpool docks. Coal from south Lancashire on the opposite bank of the Mersey could be transported cheaply by barge. The salt industry, the coal industry, and the port of Liverpool fed off of each other and together grew prosperous.

Unfortunately for Scotland, Cheshire achieved its new position of power and influence just in time to affect Scotland's union with England in 1707. After James II, a Catholic, was deposed in England, Presbyterianism was guaranteed for Scotland and the last obstacle to unification was removed. The Scottish and English parliaments merged. But adding Scotland meant bringing Scottish salt into England, and Cheshire merchants had added to the treaty of union numerous stipulations on salt production and pricing aimed at preventing Scottish salt from competing with Cheshire. This was one of several reasons the union had an acrimonious beginning. Almost thirty years before they were joined, John Collins had warned "that unless moderated in its customs" salt competition would breed enmity between England and Scotland.

Meanwhile, Cheshire salt makers would not give up on the idea that the coal beds around them extended to their region. As late as 1899, they drilled a shaft a mile deep. But again, they found only salt.

⌒

EVEN BEFORE TRUE industrialization had overtaken England, the industrial degradation of the environment was an accepted way of life in Cheshire. Cheshire merchants would look with pride at the sky, blackened twenty-four hours a day from clouds of smoke from the salt pan furnaces, and note the industriousness of their region.

The forests of Cheshire had been chopped down to fuel furnaces. Barren white scars were etched into the pastureland, where the pan scale, the residue that had to be periodically chipped off the salt pans, was dumped. And the earth itself was beginning to collapse.

In 1533, it was reported that the land near Combermere, Cheshire, had fallen in, creating a pit that filled with saltwater. In 1657, another little salt pond appeared at Bickley. In 1713, a hole appeared just south of Winsford in a place called Weaver Hall. All of these funnel-shaped holes were in proximity to salt production, and they all immediately filled with brine. Many locals believed that the holes were the result of abandoned salt mines collapsing. But mining interests pointed out that the sinkholes were not appearing near abandoned shafts. By the last two decades of the eighteenth century, when a new hole sunk every year or two, it started to appear that there was a relationship between the increasing quantities of salt being produced and the collapsing of the earth.

DESPITE CHESHIRE'S GROWING production, England still had that same dangerous dependence on foreign salt that had worried Queen Elizabeth. During the seventeenth and eighteenth centuries, it was a recurring topic of concern, especially since much of the foreign salt came from England's principal enemy, France. On land campaigns, each British soldier received a huge ration of salt so that he could acquire fresh meat along his march and salt it to use as needed. The British navy was provisioned with salt and salt foods. Salt was strategic, like gunpowder, which was also made from salt.

In 1746, Thomas Lowndes, a Cheshire native, wrote a book-length report to the admiralty on developing Britain's own sea salt supply. After studying French and Dutch salt, he, with great excitement, announced that he had discovered the secret of superior salt:

This is the Process

Let a Cheshire Salt-pan (which commonly contains about eight hundred gallons) be filled with Brine, to within about an inch of the top; then make and light the fire; and when the Brine is just lukewarm, put in about an ounce of blood from the butcher's, or the whites of two eggs: let the pan boil with all possible violence; as the scum rises take it off; when the fresh or watery part is pretty well decreased, throw into the pan the third part of a pint of new ale; or that quantity of bottoms of malt drink: upon the Brine's beginning to grain, throw into it the quantity of a small nutmeg of fresh butter; and when the liquor has salted for about half an hour, that is, has produced a good deal of salt, draw the pan, in other words, take out the salt. By this time the fire will be greatly abated, and so will the heart of the liquor. Let no more fuel be thrown on the fire; but let the brine gently cool, till one can just bear to put one's hand into it: keep the brine of that heat as near as possible; and when it has worked for some time, and is beginning to grain, throw in the quantity of a small nutmeg of fresh butter, and about two minutes after that, scatter threw the pan, as equally as may be, an ounce and three quarters of clean common allom pulverized very fine; and then instantly, with the common iron-scrape-pan, stir the brine very briskly in every part of the pan, for about a minute: then let the pan settle, and constantly feed the fire, so that the brine may never be quite scalding hot, nor near so cold as lukewarm: let the pan stand working thus, for about three days and nights, and then draw it.

The brine remaining will by this time be so cold, that it will not work at all; therefore fresh coals must be thrown upon the fire, and the brine must boil for about half an hour, but not near so violently as before the first drawing: then, with the usual instrument, take out such salt as is beginning to fall, (as they term it) and out it apart; now let the pan settle and cool. When the brine becomes no hotter than one can just bear to put one's hand into it, proceed in all respects as before; only let the quantity of allom not exceed an ounce and a quarter. And in about eight and forty hours after draw the pan.—*Thomas Lowndes, Brine-salt Improved, or The Method of Making Salt from Brine, That Shall Be as Good or Better Than French Bay-salt, 1746*

Lowndes assured the admiralty, "The greater the quantity is of salt made my way, the more satisfied the public will be, that my secret is truly made known." But, understandably, some found his secret to be excessive for just making salt, a product whose commercial success depended on low production costs. Two years later, the physician William Brownrigg in a widely distributed work, *The Art of Making Common Salt,* criticized Lowndes' formula, writing that "a purer and stronger salt can be made, and at less expense."

Cheshire salt needed to be not only better but, even more important, cheaper. Almost seventy-five years earlier, Dr. Thomas Rastel of Droitwich had written that Droitwich had simplified salt making, eliminating the cost of blood used in Cheshire brine by making an egg white scum, a technique still used in cooking to remove impurities in meat stocks for a clear aspic:

> For clarifying we use nothing but the whites of eggs, of which we take a quarter of a white and put it into a gallon or two of brine, which being beaten with the hand, lathers as if it were soap, a small quantity of which froth put into each vat raises all scum, the white of one egg clarifying 20 bushels of salt, by which means our salt is as white as anything can be: neither has it any ill savour, as that salt has that is clarified with blood. For granulating it we use nothing for the brine is so strong itself, that unless it be often stirred, it will make salt as large grained as bay-salt.—*Dr. Thomas Rastel, 1678*

The goal was always to make bay salt, salt that resembled the sea salt of Bourgneuf Bay, because this was the salt of the fisheries. Lowndes mentioned in his treatise that he had received a letter from a Captain Masters, dated June 5, 1745, estimating that the Newfoundland cod fishery used "at least ten thousand tons" of salt annually.

An eighteenth-century English engraving of codfish being salted and dried in Newfoundland. The Granger Collection

Between 1713 and 1759, through nearly global warfare with France, England had acquired most of the codfish grounds of North America. The English were excited about their new cod potential. But even a decade before they had achieved their greatest victories, Brownrigg had warned that in order to take advantage of the acquisition of Cape Breton alone, the French end of Nova Scotia, England would have to increase its supply of salt.

North American cod seemed limitless, and the only impediments to British profits were the number of ships and fishermen, and the supply of salt.

CHAPTER TWELVE

American Salt Wars

Studying a road map of almost anywhere in North America, noting the whimsical nongeometric pattern of the secondary roads, the local roads, the map reader could reasonably assume that the towns were placed and interconnected haphazardly without any scheme or design. That is because the roads are simply widened footpaths and trails, and these trails were originally cut by animals looking for salt.

Animals get the salt they need by finding brine springs, brackish water, rock salt, any natural salt available for licking. The licks, found throughout the continent, were often a flat area of several acres of barren, whitish brown or whitish gray earth. Deep holes, almost caves, were formed by the constant licking. The lick at the end of the road, because it had a salt supply, was a suitable place for a settlement. Villages were built at the licks. A salt lick near Lake Erie had a wide road made by buffalo, and the town started there was named Buffalo, New York.

When Europeans arrived, they found a great deal of salt making in North American villages. In 1541, Spanish explorer Hernando de Soto, traveling up the Mississippi, noted: "The salt is made along a river, which, when the water goes down, leaves it upon the sand. As they cannot gather the salt without a large mixture of sand, it is thrown into certain baskets they have made for the purpose, made large at the mouth and small at the bottom. These are set in the air on a ridge pole; and water being thrown on, vessels are placed under them wherein it may fall, then being strained on the fire it is boiled away, leaving salt at the bottom."

Hunter groups that did not farm did not make salt. An exception was the Bering Strait Eskimo, who took reindeer, mountain sheep, bear, seal, walrus, and other game and boiled it in seawater to give a salty taste. Many, such as the Penobscot, Menomini, and Chippewa, never used salt before Europeans arrived. Jesuit missionaries in Huron country complained that there was no salt, though one missionary suggested that Hurons had better eyesight than the French and attributed this to abstinence from wine, salt, and "other things capable of drying up the humors of the eye and impairing its tone."

The Puget Sound Indians, whose diet was largely salmon, were said to eat no salt. The Mohegan of Connecticut ate great quantities of lobster, clams, shad, lamprey, and also corn, but, according to Cotton Mather, "They had not a grain of salt in the world until we bestowed it on them."

But the Delaware salted their cornmeal. The Hopi boiled beans and squash with salt and served jackrabbit that was stewed with chili peppers and wild onions in salted water. The Zuni served boiled salted dumplings in a brine sauce and made *kushewe,* a salty bread of lime and salted suet. When a Zuni traveled, he always carried a jar or earthen box of salt along with one of red chile, a blend that would remain a classic seasoning of the Southwest.

IN THE SEVENTH month of their year, the Aztecs observed ceremonies for Vixtociatl, who was banished to the saltwaters by her brothers the rain gods, and thus she was the discoverer of salt, the inventor of salt making. The sixteenth century Spanish friar Bernardino de Sahagun described her appearance: ears of gold, yellow clothes, an iridescent green plumage, and a fishnet skirt. She carried a shield trimmed with eagle, parrot, and quetzal feathers, and she beat time with a cane topped by incense-filled paper flowers. The girl chosen to represent Vixtociatl danced for ten days with women who had made salt. Finally, on the festival day, two slaves were killed, and then the girl too was sacrificed.

Many indigenous North American cultures have a salt deity, almost always female. For the Navajo, it is an elderly woman. Among agricultural people of the U.S. Southwest and Mexico, expeditions to gather salt were often initiated with great ceremonies. Among the Hopi, this included copulation with a woman designated "the salt woman." Among many southwestern groups, salt gathering was organized by religious leaders. Usually, participants had to be initiated into a cult of salt gatherers. Often only members of a privileged clan, such as the Laguna's parrot clan, could go on salt expedition. In most cultures only men were allowed to gather salt, but the Navajo allowed women also. The Zuni, according to legend, originally allowed both, but their frivolity on the mission offended the salt goddess and the salt supply started to vanish. So they changed the custom to men only. The entire Zuni population prayed for the safe return of the salt expedition. When the men returned, the pa-

ternal aunt of each salt gatherer would wash his head and body
with yucca suds.

❧

THE HISTORY OF the Americas is one of constant warfare over
salt. Whoever controlled salt was in power. This was true before
Europeans arrived, and it continued to be the reality until after
the American Civil War.

As on the Italian peninsula, all the great centers of civilization
on the American continents were founded in places with access
to salt. The Incas were salt producers, with salt wells just outside
Cuzco. In Colombia, nomadic tribesmen probably first built permanent settlements because they needed salt and learned how to
make it. Their society was organized around natural brine springs.
The Chibcha, a highland tribe living in the area that was to become the modern capital of Bogotá, became a dominant group
because they were the best salt makers. In yet another example of
the association between sex and salt for twentieth-century psy-
chologists to ponder, the Chibcha salt lords honored the gods two
times a year by abstaining from sex and salt.

As in Africa, the Chibchas made salt by evaporating brine into
cone shapes. Befitting a multiclass society, various grades of salt
were produced, from the whitest for the rich to a black, unpleas-
ant-tasting salt for the poor. All the natural brine springs of the
Chibcha were owned by the monarch, the zipa, who ruled by
virtue of his ability to distribute salt. When the Spanish came,
having an understanding of the power of kings, they took over the
brine springs and declared them property of their king, thus destroying the authority of the zipas.

According to Bernal Díaz, the chronicler of Hernán Cortés's
conquest, the Aztecs made salt from evaporating urine. A tribe in

Honduras plunged hot sticks into the ocean and scraped off the salt, just as the Romans had observed the Britons doing. More commonly, brine from natural springs was evaporated, or desert salt beds were scraped like the sebkhas of the Sahara, or sea salt was raked from the ocean's edges.

The Aztecs controlled the salt routes by military power and were able to deny their enemies, such as Tlxalacaltecas, access to salt. William Prescott's 1819 classic, *History of the Conquest of Mexico,* described the Aztecs receiving tribute from their subjects: "2000 loaves of very white salt, refined in the shape of a mold, for the consumption only of the lords of Mexico."

The Spanish took power by taking over the saltworks of the indigenous people they conquered. Cortés, who came from southern Spain, not far from both Spanish and Portuguese saltworks, understood the power and politics of salt. He observed with admiration how the Tlatoque had maintained their independence and avoided the oppression of the Aztecs by abstaining from salt. "They ate no salt because there was none in their land," he wrote, and like the British, they feared salt dependency.

〜

THE EARLIEST EVIDENCE that has been found of Mayan salt production is dated at about 1000 B.C., but remains of earlier saltworks have been found in non-Mayan Mexico such as Oaxaca. It may be an exaggeration to claim that the great Mayan civilization rose and fell over salt. However, it rose by controlling salt production and prospered on the ability to trade salt, flourishing in spite of constant warfare over control of salt sources. By the time Europeans arrived, the civilization was in a state of decline, and one of the prime indicators of this was a breakdown in its salt trade.

The Mayan world extended from Yucatán to the present-day

Mexican state of Chiapas and across Guatemala. When Hernán Cortés first went to the Yucatán peninsula in the early sixteenth century, he found a Mayan people with a large salt industry and an extensive trade not only in salt but in salted goods such as salt fish and cured hides.

The Mayans used salt as medicine mixed with marjoram and xul tree leaves for birth control, with oil for epilepsy, with honey to lessen childbirth pains. It was also used in rituals associated with both birth and death.

In the Yucatán, salt was made from solar evaporation at least 2,000 years ago, meaning that indigenous Americans have been making solar-evaporated sea salt for at least as long as have Europeans. The Mayans also knew how to extract salt from plants, although plant salt is usually potassium chloride rather than sodium chloride. They would burn plants, certain types of palms as well as grasses, and soak their ashes into a brine that was then evaporated. This technique was practiced by isolated forest people throughout the Americas and in Africa.

The Lacandon of Chiapas are an isolated and culturally distinct Mayan group who lived self-sufficiently in a rain forest that, unfortunately for them, became the Mexican-Guatemalan border. They made salt from burning a certain species of palm, and they used this salt as money. Dressing in long white gowns, the Lacandon paddled canoes in their rain forest and lived an undisturbed and unique way of life until the twentieth century, when the modern Mexican and Guatemalan states became concerned about the international border running through the Lacandon forest. For the military, the forest made the border more difficult to guard. For some Lacandons the forest was a source of wealth and they sold the hardwood trees to lumber companies. The tribe began losing its traditions and self-sufficiency as their forest disappeared. Because the logging companies supplied them with salt, Lacandons stopped burning palms.

Typical of the cultural destruction of Chiapas Mayans, the town of La Concordia and its surrounding saltworks were flooded by a dam in the 1970s and now rest on the bottom of a lake. According to Frans Blom, the Danish anthropologist who explored Mayan culture in the 1920s through the 1940s, the site contained the unique saltworks of the Mayan highlands, where brine was diverted from springs by the use of tree trunks into shallow stone pans for solar evaporation, similar to the Hawaiian technique using stone bowls.

The people of La Concordia placed reeds in the evaporation pans, often shaped into six-pointed stars. Crystals would form on the reeds, making thick, sparkling white ornaments that salt makers sold to be used as religious offerings. By Blom's time, the Mayans were bringing these offerings to Catholic churches.

By coincidence, Cheshire salt workers had a similar tradition. At Christmastime, they placed branches in evaporation pans until salt crystalized like a fresh snowfall, and they brought these snowy branches home for Christmas decorations.

THE ARRIVAL OF the Spanish meant not only a new power controlling the salt but a huge increase in demand for industrial salt. The Spanish introduced herds of cattle that needed to be fed salt and whose hides were cured with salt in a prosperous leather industry. Obsessed with the extraction of precious metals, the Spanish invented the patio process for silver mining in mid-sixteenth-century Mexico. In this process, silver was separated from ore by using salt because the sodium in the salt extracted impurities. Silver mining by the patio process required huge quantities of salt, and the Spanish built large-scale saltworks adjacent to silver mines.

The Yucatán peninsula has a climate particularly well suited for salt production and geographically is particularly well suited for trade, with its proximity to the Caribbean and Central America. It was the largest salt producer in pre-Columbian America and remained a leader when the Spanish took it over.

The Spanish, unable to locate precious metal deposits in the Yucatán, began to look for ways of earning state revenue from the Yucatán saltworks. The Spanish Crown proposed various salt taxes. But this made the salt there more expensive, and it could not compete in Cuba with British salt. Cuba, a Spanish colony, should have been a Spanish market. But a time came in the nineteenth century, with the wild fluctuation of salt prices, when Yucatán salt was imported to England through the port of Liverpool.

THE BRITISH FIRST arrived in North America in the north, at Newfoundland, and they took cod. They next arrived in the south, the Caribbean, where they took salt, which they needed for the cod. Only after they had a significant population of colonists in between did they think of America as a market in which to sell Liverpool salt.

To the British admiralty, the solution to a lack of sea salt was to acquire through war or diplomacy places that could produce it. Portugal had both sea salt and an important fishing fleet, but needed protection, especially from the French who were regularly seizing their fishing boats. And so England and Portugal formed an alliance trading naval protection for sea salt.

The Portuguese alliance gained England access to the Cape Verde Islands, where British ships could fill their holds with sea salt on their way across the Atlantic. The islands on the eastern side of the archipelago, Maio, Boa Vista, and Sal, which means "salt," had marshes with strong brine, and in the seventeenth century Portugal granted the British exclusive use of the salt marshes of Maio and Boa Vista.

British ships had only from November until July to make salt before the summer rains ruined the brine. They usually stopped off in January, anchoring off Maio, which they called May Island. From there the sailors would row their launches less than 200 yards to a broad beach. Behind the beach was a salt marsh, where a mile-long stretch of ponds would be eight inches deep in brine. It could take months before the sailors had scraped enough salt crystals for the ship to be full. Sometimes early rains would force them to leave. Some ships had to go to Boa Vista because they found too many crews already working at Maio. At Boa Vista the brine was weaker and took longer to crystalize and the anchorage was farther out, forcing the sailors to row a mile to bring salt to the ship.

But sea salt was valuable enough for a shipload of it to be worth the labor of an entire ship and crew for several months.

⌒

IN THE SEVENTEENTH and eighteenth centuries, while European powers were fighting bitterly for Caribbean islands on which to grow sugarcane, northern Europeans—the English, the Dutch, the Swedish, and the Danes—also looked for islands with inland salt marshes like the Cape Verde Islands.

In 1568, under William of Orange, the Dutch began an eighty-year independence struggle against Spain, which cut them off from Spanish salt. But in the Americas, the Dutch could come ashore unobserved on the coast of Venezuela at Araya, a hot and desolate eighty-mile lagoon, and steal Spanish salt from the beach, where

Caribbean seawater evaporated into a thick white crust. The Dutch also got salt from Bonaire in the nearby Dutch Antilles.

The British gathered salt illegally from the Spanish on another small island in the area called Tortuga or Salt Tortuga, which is today part of Venezuela. They also made salt on Anguilla and the Turks Islands, which had the advantage of being closer to North America, where the cod fishery was. They would stop off in one of the salt islands, and, as in the Cape Verdes, the sailors themselves would scrape up salt and load up their ships and sail on to New England, Nova Scotia, or Newfoundland.

Fearing enemy warships and pirates, the salt ships traveled in convoys. They also did this in Europe for the same reasons. Huge armed fleets of ships of various nationalities would anchor off Le Croisic while salt was being loaded. Sailors were not allowed to be armed when they came ashore, because if convoys of two na-

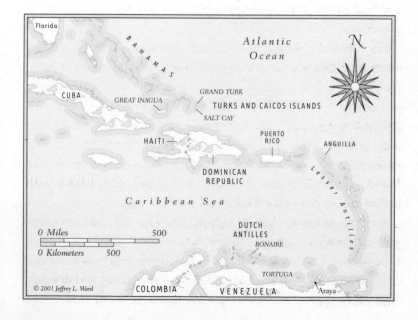

tions arrived at the same time, a port scuffle could turn into a land war. English and Dutch sailors were especially hostile toward each other.

At the end of winter, fleets of several dozen British ships, accompanied by warships, would meet in Barbados. There they would combine into one large fleet and choose a commander. Then they would go to one of the salt islands, usually Tortuga, and the crews would work for months to load their ships. If the fleet was too large or if it was a wet year, there would not be enough salt to fill all the ships, and since they were only together as a temporary arrangement, they would compete, working as fast as possible, each ship trying to secure a full hold. Then they would sail north together, and when they believed they were out of danger, especially from the Spanish fleet, each would veer off on its own course.

IN 1684, WHEN Bermuda, first explored by the British more than 150 years earlier, finally became a British colony, the first governor was given instructions to "proceed to rake salt." English ships sailing to American colonies could stop off in this cluster of minuscule islands in the middle of the Atlantic some 600 miles from the coast of North America, and pick up salt for the fisheries. It was a chance to make Bermuda productive.

But the climate in Bermuda was not warm and sunny enough for a successful sea salt operation. What Bermuda did have was cedar. So Bermudians, most of whom were originally sailors from Devon, built small, fast sloops out of cedar. Until the early eighteenth century, when New England fishermen invented the schooner, the Bermuda sloop with its single mast and enormous spread of sails was considered the fastest and best vessel under sail, capable of outrunning any naval ship. These sloops dominated the trade between the Caribbean and North American colonies and were even used in the Liverpool-to-West Africa trade.

In the Caribbean, the leading cargo carried to North America—more tonnage than even sugar, molasses, or rum—was salt. The leading return cargo from North America to the Caribbean was salt cod, used to feed slaves on sugar plantations.

In the southern part of the Bahamas chain and the group south of it known as the Turks and Caicos, salt rakers found small islands with brackish lakes in the interior. Great Inagua, Turk, South Caicos, and Salt Cay (pronounced KEY) had salty inland lakes well suited for salt making. Since Columbus and his Spanish successors had already annihilated the indigenous population, these scarcely inhabited islands were easily converted into salt centers.

Great Inagua was first raked by the Spanish and Dutch. After the Spanish killed the few local tribesmen, it was an uninhabited island and sailors from various nations stopped by and filled their ships. The Spanish named it Enagua, meaning "in water." In 1803, salt rakers from Bermuda built a small town, Matthew Town, by the edge of a salt pond at one end of the flat grassy island.

First came salt rakers, who simply scraped up what had evaporated on the edge of ponds. The crew would be dropped off on the island, where they would spend a few months or sometimes as long as a year gathering salt while the captain and three or four slaves would sail off, fishing for sea turtle, scavenging shipwrecks, and trading with pirates or between islands. Sometimes they would hide in coves by treacherous rocks or uncharted shoals and wait, or even lure the ships onto the rocks so that they could scavenge them.

An eighteenth-century Bermuda governor complained that "the Caicos trade did not fail to make its devotees somewhat ferocious, for the opportunities were in picking, plundering and wrecking." He was also concerned with the practice of sending the slaves plundering while the free sailors were gathering salt.

The governor wrote, "The Negroes learned to be public as well as private thieves."

When the captain and his slaves had finished with their profitable adventures months later, they would return and pick up their crew and a full hold of salt to sell in the North American colonies.

In the 1650s, British colonists from Bermuda sailed down to Grand Turk, a small desert island, and Salt Cay, its tiny neighbor, only two miles long by one and a half miles wide. In Salt Cay, passing ships would stop to rake the ponds that occupied one third of the island. In the 1660s, Bermudians began exploiting it more systematically, at first only in the summers, which were dry.

By 1673, the arrival of Bermudian rakers on Salt Cay was a regular event. Five years later, salt raking had become equally well organized on the slightly larger island to the north, Turk or Grand Turk Island, which was named after a native cactus thought to resemble a Turkish turban. But the Spanish would come in the winter and take the salt rakers' tools and destroy their sheds. By the early 1700s, Bermudians started living full-time on Salt Cay to protect their property. No one knows when the small harbor was built with its stone piers, but it was the most stormproof in the Turks and Caicos, a safe shelter for ships to spend a few weeks loading. But as vessels became larger, the little harbor was too shallow, and light ships had to be used to carry the salt out to mother ships anchored offshore.

Salt Cay salt makers built a system of ponds and sluices. Every year, they had to spend weeks overhauling the system. The ponds had to be drained in order to mend the stone or clay bottoms so that they would hold water and not crumble into the salt. Then the ponds would be refilled for the slow process of solar evaporation.

Salt makers came from Bermuda and built large stone houses in the Bermudan style with thick walls to hold up a cut stone roof built in steps like a pyramid. The heavy roof was designed to resist hurricanes. Mahogany furniture was brought to the island. These were the manor houses of slave plantations, but they had none of the elegance of the Virginia tobacco, or Alabama cotton, or West Indian sugar planters' homes.

The salt makers' house had an eastern porch that looked out over his salt ponds and a western porch that looked over his loading docks. The houses were always built at the water's edge by a loading dock. The salt, too valuable to entrust to anyone else, was kept in the basement, which was one story below ground, but the windowless first story of the house had no floor, so that actually each house had a two-story storage bin at its base. Salt was the salt makers' wealth, and they watched over it day and night.

Windmills pumped the seawater through successive ponds, and the mills and sluices were maintained by a blacksmith shop at the house. Slaves grew some vegetables in gardens, but the soil became poorer and poorer as the trees were all chopped down to fuel salt pans. The island was hot and dry and naked, and food and even fresh water were becoming increasingly scarce.

In 1790, a man named Stubbs who had left the North American colonies because he remained loyal to British rule, sent for his brother, Thomas Stubbs, and settled in Providenciales, an island in the Turks and Caicos. The Stubbs family were salt producers in Cheshire, but Thomas and his brother wanted to start a new life as West Indian planters. They called their plantation Cheshire Hall and tried to grow sisal, a hemp substitute from the fibers of the agave plant. But sisal growing at Cheshire Hall was a failure. Then they tried to grow sea island cotton, but that failed also. On these flat, arid little islands, everything failed except salt. Salt makers brought in livestock: donkeys to haul carts of salt to the wharves, and cattle to feed themselves as they made salt.

All that these small, salt-making islands had was their location in shipping lanes, sunshine, and marshes that trapped seawater. Yet for a time they prospered because the British Empire needed salt.

CHAPTER THIRTEEN

Salt and Independence

THE ENGLISH, THE Dutch, and the French hunted for salt, the magic elixir that could turn their new American seas of limitless fish into limitless wealth. The Dutch gave incentives to colonists and, in 1660, granted a colonist the right to build saltworks on a small island near New Amsterdam, known as Coney Island. The French learned from the indigenous people the location of licks, springs, and marshes. They used many existing saltworks, including those at Onondaga, New York, and Shawneetown, Illinois.

In 1614, Captain John Smith had explored by sea the coast of New England from Penobscot Bay to Cape Cod. Smith, one of the 105 original settlers of Jamestown and a leading force in the English settlement of North America, had also charted Virginia and the Chesapeake Bay. He did this work, both in Virginia and New England, with the intention of enticing settlers, noting the

A portrait of John Smith from a 1616 map of New England based on Smith's notes. The Houghton Library, Harvard University

prospects for enrichment from fish, salt, fruit, precious metals, furs, and even the possibility of producing silk.

Although known for a swaggering boastful nature, Smith described the riches of these new lands with notable restraint. By the early seventeenth century, a considerable literature on the wealth of the Americas had already accumulated, and most of it was outrageously exaggerated to the point where, Smith observed, settlers would quickly leave in disappointment. And so he resolved to be a realist, though his trademark style could be seen in naming Cape Ann after a woman he had been fond of during his military service in Turkey. Back in England, it was renamed Cape Ann after Prince Charles's mother.

Although personally disliking fishing, Smith understood that if it was a profitable endeavor, it would attract settlers. "Herring, cod, and ling is that triplicitie that makes their wealth and shippings multiplicities such as it is," he wrote in his unmusical prose and devoted several pages in his *Description of New England* to describing the wealth various nations had garnered from these fish. He demonstrated his point with characteristic flair, by ordering his crew to fish and salt cod while he was exploring coastlines, then earning a modest—but widely reported—fortune selling the salt cod in England and Spain.

Smith also understood the importance of salt to his dream of a British America. He had established saltworks in Jamestown in 1607. As he sailed the rocky coastline of New England, he noted places that seemed suitable harbors, but also locations that seemed favorable for salt making. He thought conditions were suitable for "white on white," reevaporating sea salt the way the English improved French bay salt. He thought that Plum Island, just north of Cape Ann, would be a particularly good site for a saltworks. In his list of twenty-five "excellent good harbors" for fishing, he completely ignored the best harbor on Cape Ann, which only nine years later would become the fishing station of Gloucester and eventually the leading cod port of New England.

Smith's *Description of New England* was an important factor in the Pilgrims' decision to go to New England, and when they arrived, they found Smith's portrait to be accurate. They were, as promised, in a land of cod where salt could be made. Though they accepted the royal name Cape Ann, they used Smith's name for Cape Cod, a name originated by his fellow Jamestown founder, Bartholomew Gosnold, because they intended, like Smith, to amass wealth from fishing. In 1630, the Reverend Francis Higgenson wrote in *New England Plantation,* "There is probabilitie that the Countrey is of an excellent temper for the making of salt." But the Pilgrims had no idea how to make it, and for that matter, they didn't know how to catch fish either.

Governor William Bradford of Plymouth Colony sent to England for advisers on fishing, salt making, and ship building. Within a few years the colony began to thrive on fishing. But it was still limited by salt supply. The salt adviser tried to make bay salt in the French manner, digging evaporation ponds lined with clay. But New England weather was ill suited to this technique. According to Bradford, the salt maker was "an ignorant, foolish, self-willed fellow."

Massachusetts, like Queen Elizabeth, encouraged salt making through the granting of monopolies to those who showed the skill to produce salt cheaply. The colony granted Samuel Winslow a ten-year monopoly to employ his ideas on salt producing, which is considered the first patent issued in America. The same year, John Jenny was given exclusive salt-making rights in Plymouth for twenty-one years. Saltworks were started in Salem, Salisbury, and Gloucester. Salt was needed not only for fish exports but also for furs. The settlers traded with the native Americans for bear, beaver, moose, and otter pelts, for which there was a lucrative European market. Because furs were salted, they were frequently exported on the same ships as cod. But to get the indigenous people to produce more furs, the British had to supply them with more salt.

The New England household also needed a great deal of salt for domestic purposes. The typical colonial New England house—the New England saltbox—got its name from being shaped like the salt containers that were in every home. New Englanders slaughtered their meat in the fall and salted it. They ate New England boiled dinner, which was either salt cod or salt beef with cabbage and turnips. They also ate a great deal of salted herring, though they seem to have preferred lightly salted and smoked red herring, perhaps because of their limited salt supply. When these early settlers hunted, they would leave red herring along their trail because the strong smell would confuse wolves, which is the origin of the expression *red herring,* meaning "a false trail."

Wealthy Virginians imported enormous amounts of English salt beef in spite of raising their own cattle. They regarded the British beef as better cured, perhaps because the British had ample salt. But Virginians made some salt of their own and imported more from England. They built a cottage industry of salted pork fat, and by the time of the American Revolution, Virginia hams were famous and exported not only to other colonies from New York to Jamaica, but even to England.

During the Revolution, when it was a part of the provisions of the Continental army, Virginia ham earned the admiration of the French, which is always considered high praise for a ham. In 1781, the Comte de Rochambeau, a French hero of the American Revolution and later tyrant of the Haitian Revolution, while engaged in the Virginia campaign said that French ham "cannot be compared to the quality and taste of theirs."

This undated recipe from Charlotte County, Virginia, is believed to have been used by the Jefferson family at Monticello.

BAKED SPICED HAM

Select a nicely cured Ham. Soak overnight in cold Water. Wipe off and put on in enough water to cover. Simmer for three Hours. Let cool in the Water it was cooked in. Take out and trim. Put into baking-pan, stick with Cloves and cover with brown Sugar. Bake in moderate Oven for two Hours. Baste with white Wine. Serve with a savoury Salad.

FOR A WHILE, the American colonists pursued their own salt making with characteristic self-reliance, producing a significant amount. But the securing of these colonies by the British had coincided with the discovery of rock salt in Cheshire and its increased production. In time, the British made Liverpool salt cheaper and more accessible than local salt, and domestic American production dropped off. This was exactly the way colonialism was supposed to work.

While relations were intact with England, the colonists had enough salt for their domestic needs, but the salt supply was inhibiting their foreign trade. Of course, they were not supposed to be engaging in foreign trade. They were supposed to buy everything from England and sell everything to England. But the American colonists produced more, especially more salt cod, than the British could sell. As long as the Americans were making their products with British salt, the British were happy to let them overproduce.

But the British often failed to supply enough salt for American needs. In 1688, Daniel Coxe wrote about New Jersey that fish were abundant but the colony was unable to establish a successful fishery because of a "want of salt." The New Jersey colony sent to France for experts—"diverse Frenchmen skillful in making salt by the sun." This was not how colonialism was supposed to work.

The American colonies, especially the two most productive, Virginia and Massachusetts, became accustomed to selling their products around the Atlantic world. New England began by selling salted cod and salted furs but soon was selling manufactured goods, buying iron and Mediterranean products in the Basque port of Bilbao, selling cod for slaves in West Africa, slaves for molasses in the Caribbean, rum made from molasses in West Africa.

By the early 1700s, Boston merchants did not feel that they needed England anymore. In one important respect, they were wrong. Despite increasingly independent and sophisticated trans-Atlantic commerce, they still depended on England for salt. New Englanders occasionally imported salt from other countries. Ships selling cod to Bilbao would pick up salt from southern Spain at Cadiz or Portuguese salt at Lisbon. But in 1775, like exemplary British colonists, the Americans were still relying on British salt—either Cheshire salt from Liverpool or sea salt from the British colonies, especially Great Inagua, Turk Island, and Salt Cay.

Tom Paine's contention that a continent obviously could not be ruled by an island was increasingly resonating among the merchant class. In 1759, the British, sensing that American trade was leading to American independence, started imposing punitive tariffs, taxes, and other measures designed to inhibit American trade. The Americans responded angrily, and the British responded with even harsher measures. In 1775, the atmosphere was so embittered that the British thought it necessary to place 3,000 troops in rebellious Boston under Major General Thomas Gage. When these soldiers attempted to spread out into the countryside, Americans went into armed rebellion, firing on the British troops at Concord and Lexington on April 19. The Continental Congress, first called in 1774 as a protest, reconvened in May 1775 to prepare for war.

In June, while the Congress was still meeting, the rebels marched on Boston and Gage committed all but 500 of his troops to a battle on Breed's Hill above Boston Harbor. Although the British objective of holding Boston was accomplished, Gage lost more than 40 percent of his soldiers in this engagement. Incorrectly known as the Battle of Bunker Hill, this was to be the worst British loss of the war.

In the summer of 1775, the British declared the colonies in open rebellion and responded with a naval blockade, causing an immediate and serious salt shortage, not only for the fisheries but for the soldiers, horses, and medical supplies of George Washington's Continental army. In addition to the blockade, British ground forces isolated the mid-Atlantic colonies from their two sources of American salt: New England and the South. They even attacked and destroyed mid-Atlantic saltworks.

~

AFTER THE BUNKER Hill debacle, Gage was replaced as commander of British forces in America by General William Howe, an illegitimate relative of the royal family. In 1758, Howe had been elected a Member of Parliament and had opposed measures against American commerce, fearing British policy would lead to a loss of the colonies. Now he was ordered to hold them by force. In August 1776, he took Long Island and then New York City. The next year, he drove Washington from Philadelphia. At this point in the war, he had successfully cut off Washington's army from its coastal salt supply and even captured Washington's salt reserves, despite the American general's desperate dispatch warning, "Every attempt must be made to save it."

The American colonists initially responded to the British blockade by boiling sea water. But boiling used up an enormous quantity of wood to make a very small amount of salt. About 400 gallons of seawater were needed to make one bushel of salt. In the winter, families would keep an iron caldron of seawater cooking over the household fire, which was not a great additional expense, since the fire was burning continuously anyway to heat the house. But only a small amount of salt was produced this way. Salt makers drove wooden stakes into tidal pools, and salt would crystalize on the wood as the pools evaporated. This technique was inexpensive but also yielded little.

The Continental Congress passed several measures addressing the salt shortage. On December 29, 1775, the Congress "earnestly recommended to the several Assemblies and Conventions to promote by sufficient public encouragement the making of salt in their respective colonies."

In March 1776, *Pennsylvania Magazine* published a lengthy excerpt from Brownrigg's essay on making bay salt. The article was reprinted as a pamphlet and circulated by the Congress. On May 28, 1776, the Congress decided to give a bounty of one third of a dollar per bushel, which weighs about fifty pounds, to all salt importers or manufacturers in the colonies for the next year. The moment the pamphlet and bounty offer were published, saltworks were started along the American coastline. New Jersey had contemplated establishing a state-operated saltworks, but so many private ones were built along its coast in 1777 that it canceled these plans as unnecessary.

In June 1777, a congressional committee was appointed "to devise ways and means of supplying the United States with salt." Ten days later the committee proposed that each colony could offer financial incentives to both importers and producers of salt. Some of the thirteen colonies had already been doing this. New Jersey declared that any saltwork could exempt up to ten employees from military service.

~

ONE OF THE many sea salt operations to start up in response to the government's publication of the Brownrigg pamphlet and bounty offer was the first saltworks on Cape Cod. Given the cod-fishing communities and the presence of sea and wind, Cape Cod was a logical place to make salt. The water both on the bay side and in Nantucket Sound is even saltier than that of the open Atlantic.

The first works was started in the town of Dennis by John Sears, who spent his days so lost in thought that he was known as "Sleepy John Sears." The neighbors were skeptical of the 10-by-100-foot wooden vat Sleepy John built in Sesuit Harbor. The vat leaked, and after many weeks he had produced only eight bushels of salt.

The neighbors laughed, but Sleepy John Sears spent the winter caulking the vat as a ship's hull would be sealed. In the summer of 1777, a time of great salt scarcity, he produced thirty bushels of salt and his neighbors stopped laughing, and Sleepy John became known as Salty John Sears.

The following year, the British man-of-war the *Somerset* ran aground trying to round the Cape. The coastline was poorly marked, and scavenging shipwrecks was a strong local tradition. Sears took the *Somerset*'s bilge pump to fill his vats. But even with the bilge pump, producing Sears's salt required a great deal of heavy manual labor, and only wartime prices made this high-cost salt economically viable.

Then a man named Nathaniel Freeman, from nearby Harwich, suggested that Sears use windmills to pump seawater. The same thing had been done in eighth-century Sicily, in Trapani, but Cape Codders thought this was a brilliant new idea. Soon the wooden skeletons of rustic windmills were seen on the edges of most Cape Cod towns. The windmills, known as saltmills, pumped seawater through pipes—lead-lined hollowed pine logs—to the evaporation pans. But in a climate where solar evaporation was viable only in the summer months, the hardship of wartime made this operation profitable. And still these rebel colonies could not produce enough salt to meet their needs.

Fishermen with catches to be salted and farmers with pigs and cattle to slaughter and salt before winter hoped for a short war, but that was not to be. By the time it ended with the Treaty of Paris in September 1783, the American Revolution would be, until Vietnam, the longest war ever fought by the United States. A new nation was born with the bitter memory of what it meant to depend on others for salt.

CHAPTER FOURTEEN

Liberté, Egalité, Tax Breaks

I N 1875, A prominent German botanist named Matthais Jakob Schleiden wrote a book, *Das Salz,* which contended that there was a direct correlation between salt taxes and despots. He pointed out that neither ancient Athens nor Rome, while it remained a republic, taxed salt, but listed Mexico and China among the salt-taxing tyrannies of his day. It seems uncertain if salt taxes are always an accurate litmus test for democracy, but the French salt tax, the gabelle, clearly demonstrated what was wrong with the French monarchy.

The argument for the gabelle had been that since everyone, rich and poor, used salt more or less equally, a tax on salt would be in effect a poll tax, an equal tax per person. Throughout history, poll taxes, charging the same to the poorest peasant as the richest aristocrat, have been the most hated. The gabelle was not an exception. The tax performed the peculiar service of making a very common product seem rare because the complex rules of taxation inhibited trade. And even more infuriating, the gabelle made a basic product expensive, for the profit of the Crown.

Even the Crown's claim that the gabelle was fair because it taxed everyone equally was not true. There were many arbitrary provisions, such as the exemptions for the town of Collioure; for some, but not all religious institutions; some officers; some magistrates; some people of note.

The gabelle, like France, was established piecemeal. The first attempt at a comprehensive salt administration occurred in the Berre saltworks near Marseilles in 1259, by Saint Louis's brother, Comte Charles de Provence. The following century, this administration was extended to Peccais, Aigues-Mortes, and the Camargue—an area that became known administratively as Pays de Petite Gabelle. In 1341, Philip VI established a salt administration in northern France that was labeled the Pays de Grande Gabelle. At the time, these two areas included most of the territory controlled by the French Crown.

At first the gabelle imposed a modest 1.66 percent sales tax on salt. But each monarch eventually found himself in a crisis—a prince to be ransomed, a war to be declared— that was resolved by an increase in the salt tax. By 1660, King Louis XIV regarded the gabelle as a leading source of state revenues.

One of the gabelle's most irritating inventions was the *sel du devoir,* the salt duty. Every person in the Grande Gabelle over the age of eight was required to purchase seven kilograms (15.4 pounds) of salt each year at a fixed high government price. This was far more salt than could possibly be used, unless it was for making salt fish, sausages, hams, and other salt-cured goods. But using the sel du devoir to make salted products was illegal, and, if caught, the perpetrator would be charged with the crime of *faux saunage,* salt fraud, which carried severe penalties. Many simple acts were grounds for a charge of faux saunage. In the Camargue, shepherds who let their flocks drink the salty pond water could be charged with avoiding the gabelle.

A 1670 revision of the criminal code found yet another use for salt in France. To enforce the law against suicide, it was ordered that the bodies of people who took their own lives be salted, brought before a judge, and sentenced to public display. Nor could the accused escape their day in court by dying in the often miserable conditions of the prisons. They too would be salted and put on trial. Breton historians have discovered that in 1784 in the

A woodcut showing salt being measured, from Ordonnances de la Prévôte des Marchands de Paris *1500*. The Granger Collection

town of Cornouaille, Maurice LeCorre had died in prison and was ordered salted for trial. But due to some bureaucratic error, the corpse did not get a trial date and was found by a prison guard more than seven years later, not only salted but fermented in beer, at which point it was buried without trial.

⌒

LOUIS XIV PUT the state's finance and commerce in the hands of Jean-Baptiste Colbert, the son of a merchant family from the Champagne region. Colbert was a leading advocate of the school of economics known as mercantilism, which held that the value of the state was measured in the goods it exported and the precious metals it imported. To this end, both production and trade were to be tightly controlled by the state through such tools as taxes and tariffs. Mercantilism held that the sum total of world trade was limited, so that if England increased its trade, it would be at the expense of France and everyone else.

In salt, Colbert reasoned, France had a valuable export product. He was directly involved in marketing French salt to the Nordic countries, and he made important improvements in French waterways to move salt more efficiently. He believed that France made the world's best salt, which is not a surprising point of view for a Frenchman, but many Englishmen, Dutch, and Germans of the time agreed. Colbert corresponded with salt makers about technical improvements. Today, in Guérande it is proudly asserted that Louis XIV would only eat Guérande salt. If this is true, it would explain why he was so interested in improving the color. Colbert pointed out that if it were white like the salt of France's primary salt competitors, Spain and Portugal, it would sell better. But the paludiers of Guérande continued to rake up salt with green algae and the charcoal-colored clay mud on which the ponds were built.

Colbert's name became infamous in French salt history when, in 1680, he revised the gabelle, codifying the inequities among regions into six unequal zones. The Pays de Grand Gabelle was the oldest part, the heart of France, including the Paris region. With only one third of the French population, who used only a quarter of the French salt, and yet who paid two-thirds of the state's salt revenue, the residents of this region were the angriest people in France. Local merchants imported inexpensive salt from Portugal, the white salt of Setúbal, to try to lower the local price.

The Pays de Petite Gabelle, on the Mediterranean, where much of the salt production was owned by the Crown, was less rigidly controlled but also so heavily taxed that one fourth of all salt tax revenue was squeezed from this fifth of the population.

In the third region, the Pays de Salines, which included Lorraine and other areas of inland brine wells, the Crown also owned much of the production. But in this region, unlike the Grande and Petite Gabelles, both wholesale and retail sales were carried out by private merchants rather than agents of the state. Hence, far less political tension existed in this region, but far less revenue was earned by the Crown. People in the Pays de Salines consumed twice as much salt as those in the Grande Gabelle. Adam Smith would no doubt have argued that the relative free trade of the Pays de Salines increased sales, though it could also be argued that the northern French did not traditionally eat as much salted food as in eastern France with its ham, sausage, and choucroute.

In southwestern France, François I, the sixteenth-century monarch who kept his table salt in a Cellini, dropped the small but irritating consumer tax and replaced it with a much larger tax on producers. After a year of angry protest, he cut this tax in half, and a year later, in 1543, the tax was entirely dropped. Instead, the rigid administration of the Grande Gabelle, with its controls on production, wholesale operations, and retail sales, was to be extended to this region. The result was a movement of some 40,000 farmers, who rose up in armed rebellion with the slogan "Vive le roi sans gabelle"—Long live the king without the gabelle. The French Crown, shocked by the size and ferocity of the uprising, backed down, thankful that at least they were still saying "Vive le roi."

It was decided that in this troubled region, known to the gabelle as Pays Redimées, it would be prudent for the Crown to content itself with only the revenue from tolls on the transport of salt. When Colbert codified the gabelle, he kept this southwestern Pays Redimées with its exemptions, which meant that while the north endured severe salt controls, and the east and the Mediterranean also had some, untaxed southwestern salt could be traded across the region's southern border into Spain at competitive prices. Worse, the Basque provinces, which were right on the Spanish border, as a condition of their participation in France, were exempted from the gabelle, leaving the Basques in an ideal position to trade inexpensive salt in both countries.

Added to the gabelle's complicated multitiered system was yet another privileged border region, a small area on the English Channel where salt makers boiled seawater with the ashes of seaweed, making a fine, white salt similar to peat salt. The Crown had pleased this region with only light salt taxes, but then realizing that such high-quality inexpensive salt could easily flood the neighboring Grande Gabelle, restricted the amount of salt produced by limiting the number of saltworks.

To the rest of France, the most irritatingly unequal region was the Pays Exempt, made up of Brittany and newly acquired areas in the north such as Flanders, which had been brought into

France with the promise that they would not have to participate in the gabelle. These were also fishing regions, like Collioure, and Colbert wanted fishing areas to be exempt from the salt tax. He had a great belief in the value not only of salt fish as an export but of fishermen as a potential navy.

The Pays Exempt also included the entire salt-producing coast from Guérande to the salt-making islands off La Rochelle. In the mid–eighteenth century, salt provided work for 950 families in the Guérande region alone. About 500 men were paludiers, working 32,000 salt ponds. More than 3,000 ponds had been added since the sixteenth century. The gabelle had made paludiers an agricultural elite, earning a better living than most peasants of the time. Some even owned small parts of their saltworks.

Brittany, like Basque country, was a border region with cheap untaxed salt to trade, in this case across the Channel to England and up the coast to Holland. In the seventeenth century, so many English, Welsh, Scottish, and Dutch ships put into Le Croisic for salt that the local Catholic church worried about a Protestant influence on the townspeople.

⤙

IN 1784, THE French government turned to Jacques Necker, a Swiss banker so brilliant in his administration of the disastrous French economy that for a moment it appeared he would save the monarchy. In 1784, he reported that a *minot* of salt, which was forty-nine kilograms (107.8 pounds), cost only 31 sous in Brittany, but 81 in Poitou, 591 in Anjou, and 611 in Berry. Necker recognized that with such price differences, France was rich in opportunities for smugglers.

Salt smugglers and clandestine salt makers, the *faux-sauniers,* were simply opportunists amassing illegal fortunes underselling legal salt. Yet, they became popular heroes who could wander the countryside helping themselves to farm products without ever hearing a complaint from a peasant. Colbert's 1680 revision of the gabelle made it a crime for an innkeeper to give a room to a salt smuggler. A repeat offender could be sentenced to death.

The *gabelous,* the hated collectors and enforcers of the gabelle, were often crude and lawless men, abusive of their special privileges, which included carrying arms and stopping, questioning, searching, or arresting people at will. The gabelous were especially distrustful of women and abusive to them, sometimes squeezing a choice part for the pleasure of it. Often, to their disappointment, they would find bags of salt in these places. Women hid salt in their breasts, corsets, posteriors—places where they hoped not to be squeezed. Sometimes entire *faux culs,* false rears, would be constructed for hiding salt in a dress.

Something close to a state of permanent warfare developed between salt smugglers and the gabelous. Gabelous would be murdered, and the Crown would respond by having royal troops sack the village where the crime took place. On September 8, 1710, the gabelous went heavily armed into the woods near Avignon to intercept salt smugglers, and forty or fifty salt traders opened fire on them. The area was in open rebellion. Similar rebellions were breaking out all over France.

The most important smuggling border in France was the Loire River, which marked the line between the Pays Exempt and the Pays de Grande Gabelle, between Brittany and Anjou, regions where Necker had found the price of salt to be 31 sous and 591 sous, respectively. In 1698, a government official reported that "salt smuggling is endless on the Loire." Normally impoverished peasants living along the river could earn comfortable incomes from moving salt. The legendary smugglers had colorful pseudo-nyms, such as François Gantier a.k.a. Pot au Lait, milk pitcher. Because local fishermen, who knew all the hidden islands and coves of the river, would carry salt, the Crown declared it illegal to fish at night. By 1773, the gabelle had 3,000 troops stationed on the Loire to stop salt smuggling.

Some ruses were complicated. Salt cod that landed at Le Croisic was moved up the Loire for sale in France. Some of the cod had been salted and dried on land, but another product, known as green salt cod, not because of its color but because it was closer to its natural state, was made on board ship, where it was only salted but not dried. This was a more delicate cure and required ample salt to prevent spoilage while in transit. But at times, it seemed to inspectors, the fish was considerably oversalt-ed. Cod would be shipped in thick layers of salt. Salt inspectors on the Loire would examine the cod shipments from Le Croisic entering the Pays de Grande Gabelle, fish by fish, shaking off the excess salt, making note of how much salt fell off of how many fish. If too much salt fell off, it would be reported. However, mer-chants found that their journey could be expedited by the gift of a few salt cod to the right official.

BY THE LATE eighteenth century, more than 3,000 French men, women, and even children were sentenced to prison or death every year for crimes against the gabelle. The salt law in France, as would later happen in India, was not the singular cause of revolution, but it became a symbol for all the injustices of government.

In 1789, the French revolted, declaring the establishment of a National Assembly. When King Louis XVI tried to send troops against this revolutionary legislature, a mob attacked the Bastille and an armed revolution began. That same year, the revolutionary legislature repealed the gabelle. Some in the Assembly had argued for a low salt tax universally applied. But in the end the Assembly voted for no salt tax at all, not even bothering to replace this mainstay of state revenues with another source of income.

On March 22, 1790, the National Assembly, calling the salt tax "odious," annulled all trials for violation of the gabelle and ordered all those charged, on trial, or convicted to be set free.

Louis, accused of conspiring with Austrians and Prussians to overthrow the revolution, was beheaded. His wife, Marie Antoinette, who loved choucroute, was also beheaded, as were many of the Swiss soldiers of the Garde Royale. They also had acquired the court taste for choucroute and numerous inns had sprung up near the Palais Royal, where they had spent their meal breaks, feasting on choucroute with sausages and salted meats. The tradition of restaurants serving midday choucroute in that part of Paris continues to this day.

IN 1804, NAPOLÉON Bonaparte, who had risen to head of the revolutionary army and then rose to first consul, became emperor of the French. He reinstated the gabelle but without an exemption for Brittany.

Their salt no longer having a competitive advantage, the paludiers, instead of being slightly better off than the average French peasant, were now among the poorest. They continued to wear large, floppy, three-cornered hats in the style of eighteenth-century peasants. Visitors found this a picturesque part of Brittany. Novelist Honoré de Balzac abandoned poetic restraint in his description of the paludiers and their treeless salt marsh, writing that they had "the grace of a bouquet of violets" and asserting that it "was something a traveler could see nowhere else in France." He compared the area to Africa, and in the age of French colonialism many followed, comparing the impoverished Breton paludiers to Tuaregs, Arabs, and Asians. Faced with an onslaught of affluent French who found them exotic, the paludiers made souvenirs: ceramic plates depicting their dress and dolls in paludier costume fashioned out of seashells. Le Bourg de Batz became Batz-sur-Mer,

Watercolor of a paludier from an 1829 book by H. Charpentier illustrating the clothing worn by salt workers in the Guérande area. Musée des Marais Salants, Batz-sur-Mer

Batz-by-the-sea, to make this salt town by the swamp sound more suitable for tourism.

The salt cuisine of Brittany showed its poverty. Breton cooking was based on the few simple crops that paludiers could grow in their clay-bound soil, mostly potatoes and onions, which absorbed a salty taste from the seaweed in the soil. *Ragoût de berniques,* literally a stew made of nothing, was in fact made of potatoes, carrots, and onions. While France was one of the last European nations to accept the eating of potatoes, Brittany was one of the first potato-eating parts of France. Almost forty years earlier Antoine-Augustin Parmentier had persuaded the royal family to promote the eating of potatoes, a man named Blanchet launched a potato-eating campaign in Brittany. Soon after that, a cleric named de la Marche distributed potatoes to poor parishioners and was nicknamed *d'eskop ar patatez,* the potato bishop.

A nineteenth-century postcard of sardines being salted in Pouliguen, near Batz. Musée des Marais Salants, Batz-sur-Mer

After the Revolution, paludiers supplemented their diminished income by growing potatoes, which were boiled in brine that left a fine salt powder on the skin—*patate cuit au sel.*

A Breton expression was "Kement a zo fall, a gar ar sall"— Everything that is not good asks to be salted. Everything from meat to butter to potatoes was salted. Salt was Brittany's cheapest product, the one everyone could afford. Another Breton proverb was "Aviz hag holen a roer d'an nep a c'houlenn"—Advice and salt are available to anyone who wants it.

Kig-sall, salted pig, usually was made with the ears, tail, and feet—sometimes better cuts if they could be afforded—put in a barrel with lard and salt, and kept two or three months until preserved like ham. And there was *oing,* known in Breton as *bloneg,* which was nothing more than pork fat rendered with salt and pepper, dried in the open air on paper, and then smoked in a fireplace. A slice of oing was added to a vegetable soup as a substitute for meat.

⌒

IN THE 1870s, when the area was connected to the national railway system, the floppy, three-cornered hats vanished. The same railroad system favored eastern France, where the new industries such as steel were, and made the salt of Lorraine more accessible than sea salt. The gabelle remained a part of French administration until it was finally abolished in the newly liberated France of 1946.

Honoré-Gabriel Riqueti, the Comte de Mirabeau, the man who had defied Louis XVI by opening the National Assembly, said, "In the final analysis, the people will judge the revolution by this fact alone—does it take more or less money? Are they better off? Do they have more work? And is that work better paid?"

CHAPTER FIFTEEN

~

Preserving Independence

TREATIES ARE USUALLY imperfect solutions, and the
Treaty of Paris did not end all hostilities between the new
United States and Britain. The United States was embargoed
from any British goods, and British colonies were not permitted
to engage in U.S. trade. The Turk Islands, the present-day Turks
and Caicos, including Salt Cay, became havens for Americans
still loyal to Britain. In Cape Cod the price of salt rose from fifty
cents a bushel to eight dollars.

In 1793, in a postwar economy that was still demanding salt,
another Sears, Reuben Sears, a Cape Cod carpenter, invented a
roof that slid open and shut on oak rollers, allowing sea salt to now
be made efficiently from March until November. The vats were
exposed while the sun was shining, but after sunset and whenev-
er it began to rain the roofs were rolled over the vats. Though the
saltworks were privately owned, the Cape Cod communities considered them so essential to the general well-being that when

One of the last Cape Cod saltworks still in operation in Yarmouth in the late nineteenth century with vats full of brine and rolling roofs open. An operating windmill is in the background. The Snow Library, Orleans, Massachusetts

clouds began to darken the daytime sky, men and women would run out to roll all the roofs closed and children would be sent from the schools to help, and the coastline would rumble like nearby thunder from the sound of hundreds of oak wheels.

The small-scale Yankee entrepreneur, for whom New England was famous, found an opportunity in salt. By 1800, a small initial investment in a Cape Cod saltworks would quickly yield returns of 30 percent. Most of the stretches of virgin sand beach and upland dunes, land considered useless until then, were becoming marred with windmills, pipes, and huge vats with rolling roofs. The prices were high, and the market seemed endless. Whatever salt was not used by local fishermen was shipped to Boston or New York. As long as the profits were copious and easy, Cape Codders cared no more about their spoiled dunes than did the people in Cheshire worry about their blackened skies.

On Cape Cod they talked of "the lazy man's gold mine," and it seemed everybody wanted to get into salt making. The glassworks in Sandwich, famous in the nineteenth century for their little glass saltcellars, needed intense heat for glassmaking. The cooling fires still gave off enough heat to evaporate sea-water, and salt became a by-product of their glassmaking.

With an increased salt supply, the fishing industry grew.

THE AMERICANS DID not forget the salt shortages of the Revolution. Several states, including Massachusetts, still paid bounties to salt producers. The new nation remained, in principle, determined to encourage salt production. In practice, this was not always the case. When the new government realized that there was an unregulated commerce in whiskey in western Pennsylvania, traded across the Allegheny for salt, it responded by taxing the whiskey in order to stop the trade. In 1791, the whiskey-producing farmers rebelled, and beloved President Washington shocked the public by calling out a militia to put down what has become known as the Whiskey Rebellion.

In 1787, the new Americans began producing salt in Onondaga, New York. Jesuit missionaries who had been with the Onondaga tribe in the seventeenth century, first told Europeans of the salt springs there. One French missionary, Father Simon Le Moyne, wrote in his diary: "We tasted a spring which the Indians dared not drink. They say it is inhabited by a demon who makes it fetid. I found it was a salt spring. In fact, we made salt as good as sea salt and carried a quantity of it to Quebec."

The Onondaga are an Iroquois-speaking tribe, hosts to the annual meeting of the Iroquois. Like some, but not all, of the Iroquois groups, they had sided with the British during the Revolution. In 1788, New York State negotiated a treaty with the Onondaga establishing a 10,000-acre reservation with joint ownership. But in 1795, the treaty was renegotiated, and the Onondaga, a people that traditionally had not even used salt, gave up their rights to the land in exchange for the annual delivery to the Onondaga nation of 150 bushels of salt. In 1787, when the whites began producing salt there, it had been at a rate of only ten bushels a day.

The state is still delivering its annual salt payments, though some Onondaga now feel that they would rather have the land returned. The payments amount to a truckload of five-pound bags, which the state buys for the best price it can find—usually between $1,000 and $2,000, which is not a huge increase from the $900 that 150 bushels of salt cost at the time. Today the Onondaga use much of their salt for preserving deer and other game and making a great deal of sauerkraut, neither of which was a tradition before contact with Europeans. According to Audrey Shenandoah, a member of the Onondaga: "We make a lot of sauerkraut, that is, since the contact. We grow a lot of garden vegetables. A lot of cabbage and make sauerkraut. But we didn't use salt for much but medicine before the contact." Salt is still used by the Onondaga to draw out the infection from an insect bite or the prick of a thorn.

In 1797, the state of New York began granting leases for working the brine springs of Onondaga. The state fixed a maximum price of sixty cents per bushel with a four-cents-per-bushel tax. That year, production at the springs, centered in the town of Salina, was 25,500 bushels, but by 1810, Onondaga and Cayuga Counties were producing about 3 million bushels annually, using both solar and wood-fired evaporation. The brine springs had become the most important saltworks in the new United States.

THE FEELING WAS strong in the United States that the British were not to be trusted. They had never withdrawn as promised from U.S. territory along the Great Lakes, they encouraged the hostility toward the United States of native Americans, including the Onondaga and Cayuga, and they refused any agreement that would be in any way helpful to the U.S. economy. John Jay's 1795 treaty with Britain, which opened trade but on terms more favorable to England than the United States, was unpopular.

The British claimed the right to force into their own service any British sailor serving on a U.S. ship and boarded American merchant vessels in search of them. They frequently also took American sailors. In 1807, a British warship fired on the American frigate *Chesapeake*. President Thomas Jefferson responded with the Embargo Act, banning U.S. ships from foreign trade. This act was aimed at both the British and the French because both boarded American merchant vessels. Not only did the trade ban fail to change European policy, but it was an economic disaster for New England. The embargo was dropped, but there were calls for a retaliatory invasion of the remaining British colonies in North America, present-day Canada.

In a young country in which the North and South were increasingly at odds with each other, the cry for such radical measures against Canada usually came from Southerners. New Yorkers simply complained that Canada was able to attract the larger share of upstate commerce because it had better commercial waterways.

In 1808, a resolution recommending consideration of a canal connecting the Great Lakes to the Hudson River was introduced in the New York Assembly by Joshua Forman, from the salt-producing town of Salina. Forman believed that the canal was the necessary key to expanding the salt industry. It would offer the

Onondaga salt region an inexpensive route for bulk shipment to New York City. From there, the world would be their market.

Despite considerable opposition to the proposal, $600 was appropriated to survey a possible route. Politicians and financiers in New York were distrustful of the project, fearing it would undermine the importance of the port of New York. The exception was the former mayor of New York City and current governor, De Witt Clinton. The leading advocate of the canal, Governor Clinton was from one of the most prominent New York families. His father, James, had been a Revolutionary War hero, and his uncle, George, served as vice president from 1805 to 1812 under both Thomas Jefferson and James Madison. But both of these presidents expressed their doubts about the project.

Governor Clinton appointed James Geddes to make the survey. Geddes, who lived in Onondaga County, was a lawyer, a judge, a former state legislator, and an amateur surveyor. The salt-producing town of Geddes was named after him, and he had been one of the pioneers of the local salt industry—the industry that needed the canal. He spent most of 1808 traveling between the Hudson River and Lake Erie, examining the topography.

In 1809, a New York delegation went to Jefferson, hoping to persuade him to consign federal funding for the project. The moment seemed auspicious. For the first time in the short history of the United States, the nation was solvent, had settled its huge debt, and revenue was expanding. But Jefferson said, "It is a splendid project and may be executed a century hence," and concluded, "It is little short of madness to think of it at this time."

Clinton then went looking for New York State funding for the project that was increasingly known as "Clinton's ditch." In 1810, the New York state legislature approved a "Board of Commissioners" with a $3,000 budget to investigate the feasibility of constructing a commercial canal connecting Lake Erie and the Hudson River. If this could be done, the United States would have a waterway from New York City to what is now the Midwest but was then thought of as the western frontier.

The public regarded this commission as a scam—a summer vacation at taxpayers' expense to upstate New York, an area viewed in New York City as a scenic vacation ground. This suspicion was reinforced by the revelation that a number of the commissioners were planning to take their wives.

The commission was losing support and probably would have been canceled had it not been for a completely unrelated concern about a mud bar. The legislature assigned the canal commission to investigate the mud bar, which was of far greater public concern than the possible canal.

De Witt Clinton took part in the 1810 commission, which reported that the saltworks were producing far below their capacity but were limited by poor roads. Exploring alternatives, they asked locals what they felt would be the result if they built a good road from the saltworks to Lake Erie, only nine miles away. Locals gave Clinton his ideal response by agreeing that such a road would only profit the British. Canadian schooners on Lake Erie would pick up the salt and sell it in British North America.

The final argument for the canal was the inevitable war with Britain from 1812 to 1815. When this war broke out, the Americans were faced once again with a salt shortage. The British blockaded Massachusetts and tried to prevent Cape Cod salt from reaching Boston or New York, though wily New England sailors sometimes slipped through at night. In December 1814, the British landed a warship in Rock Harbor, on the bay side of Orleans, Cape Cod, and threatened to burn down the local saltworks. The summer before, the British had gotten to Washington,

D.C., and burned most of the public buildings including the presidential residence, forcing President Madison to flee. So no one in Orleans doubted the British resolve to burn their little windmill-and-rolling-roof saltworks.

A Cape Cod militia was waiting on the beach when the British attempted to land, and reportedly shot and killed two British sailors in a brief skirmish that forced the British to withdraw. A month later, Andrew Jackson won the final battle, the Battle of New Orleans, in which 2,000 British soldiers died, and Cape Codders could not resist calling their own brief engagement over their saltworks "the Battle of Orleans."

AS SOON AS the war ended, lawmakers pushed to approve the Erie Canal, and work began in 1817. The estimated cost of the project was $6 million—almost $5 per inhabitant of New York State.

Among the state's plans to finance the completion of the canal was to tax upstate salt at a rate of 12.5 cents per bushel. It was one of the few salt taxes in history that was not resented. The canal would bring prosperity to the salt region.

The canal was built in sections, and each was put into service upon its completion. The first section to be completed was the ninety-eight miles from Utica to the Seneca River, the section that ran through the salt-producing region. A weigh station to assess barge loads, designed to look like a Grecian waterside temple, was built in the tiny, backwoods stopover of Syracuse.

In October 1825, the last section of the canal was completed, and Governor Clinton and other notables went to Buffalo to board the boats of a flotilla making the inaugural voyage to New York City. The lead vessel was the *Seneca Chief*, which carried a portrait of Clinton in Roman toga by a distinguished lithographer of the day. Among the notables joining Clinton on board was Joshua Forman. The vessel was provisioned with symbolic items, including two kegs of Lake Erie water to be poured into the Atlantic off Sandy Hook, some whitefish, a canoe made by Lake Superior tribesmen, and potash from the saltworks.

The Erie Canal was 363 miles long. Like most famous bridges and waterways, it occasionally attracted unhappy romantics wishing to leap to their death. But when they hit Clinton's ditch, they were shocked to discover that, with limited funding, the state had only been able to afford to dig the canal four feet deep.

THE CANAL HAD opened at a prosperous time for American salt. In 1837, Cape Cod alone had 658 salt companies producing more than 26,000 tons per year. But Cape Cod lost its competitive advantage once upstate New York had its own waterway to New York City.

Not only did the New Yorkers now have more efficient transportation, but, borrowing Cape Cod ideas, they also made their salt more efficiently.

The Salt Springs are in the towns of Liverpool, Salina, and Syracuse in the county of Onondaga. . . . The works which we had an opportunity of examining in Syracuse, are constructed upon the plan of the works upon the Cape and on our own coasts and beaches—open and extensive vats, covered at night and during the rainy and wet weather. But evaporation is hastened by boiling the water in large kettles constructed on purpose, in Liverpool and Salina. Wood is abundant and so cheap, that the expense is very trifling, the

water is drawn up by horse and steam power, and it is estimated that 90 gallons of water will make one bushel of salt,
so perfectly is the water saturated with salt.—Barnstable Patriot, *September 4, 1830*

The New York saltworks used Cape Cod–style rolling roofs.
Apparently, something about these roofs was great fun. In both
Cape Cod and Syracuse, families were constantly complaining
about their young sons slipping away to the saltworks and wearing out their pants sliding down the roofs and crawling between

*Half of an 1890 stereo card of the Onondaga saltworks showing salt being
raked from solar evaporation vats with rolling roofs of a design copied
from the Cape Cod saltworks.* Onondaga County Salt Museum, Liverpool,
New York

the vats. Childhood memoirs contain detailed descriptions of the saltworks as playgrounds, with acres of vats dripping stalactites of white, amber, and rust red.

As the New York saltworks prospered, with salt vats taking up more and more acres, it became more difficult to roll back all the roofs in the face of a sudden downpour. Watchtowers were built with warning bells. If the rain watchers saw dark clouds, they rang the bells and hundreds of workmen and their families immediately ran to push the covers over the vats.

The workers lived in villages close around the saltworks so they could be near the vats when the bells rang. Then entire families ran to the saltworks, competing with each other to be first to cover a complete row. Winning families got small cash prizes.

Many of the New York salt workers were Irish. The Irish soaked both potatoes and corn in brine. Salt potatoes, new potatoes cooked in brine much the same as were made by the salt workers in Guérande, are still a specialty of Syracuse.

Salina was an important center with hundreds of saltworks and several hundred homes. Syracuse, chosen as the best route for canals, had been an undeveloped swampy lowland. Colonel William L. Stone, passing through in 1820 when the Syracuse population was 250 people, wrote, "It was so desolate it would make an owl weep to fly over it."

The Erie Canal ran west to east, and the Oswego Canal, which connected the Erie Canal to Lake Ontario, ran north to south. The two intersected in the center of the town of Syracuse. With its torch-lit bridges over reflecting canals, Syracuse became known as the "American Venice." Once Syracuse became Venice, Salina was reduced to a suburb. Syracuse was now, like the Italian Venice, a salt port, where Onondaga salt was loaded onto Erie Canal barges. By the time the full canal was opened, only five years after the town was said to sadden stray owls, Syracuse had tripled its population, and by 1850, 22,000 people lived there.

Not only did New York have a Venice, it had a Liverpool, the town being named so that Onondaga salt could be shipped around the United States with that trusted old brand name "Liverpool salt."

AFTER THE AMERICAN Revolution, a debate had begun about where to locate the capital of the new country. Virginians, arguing for the Potomac, made the outlandish claim that this Virginia river, which empties into Chesapeake Bay, also connected with the Ohio. This would have made the Potomac America's central waterway, since the Ohio cuts through the Midwest and enters the Mississippi. It was a Virginia propaganda ploy, and no such river connecting the mid-Atlantic to the Mississippi exists. However, Virginia did have a river that, while it never reaches the Atlantic, connected Virginia to the Ohio River. In the western part of the state, today West Virginia, the Great Kanawha River begins and flows into the Ohio, carrying goods and people to Cincinnati and Louisville. As Americans moved west across the Appalachians, this was one of the major routes. The river trade and migration turned the frontier town of Charleston, West Virginia, into a trading center. Pivotal to this trade and even more pivotal to the economic development of the Midwest was a ten-mile stretch of the Kanawha that produced salt.

On the northern bank of the Great Kanawha was a huge salt lick known as the Great Buffalo Lick. The first Europeans in the area noted indigenous people making salt at the lick and also noted the many wide trails made by buffalo and deer to this place. In fact, it was animals, not so-called trailblazers such as Daniel

Boone, that had carved the original trail across the Allegheny Mountains to the Ohio River Valley.

In 1769, when Daniel Boone followed that trail, crossing the Cumberland Gap into Kentucky, he took with him Kanawha salt, as did the thousands of other settlers that followed. In 1797, a man named Elisha Brooks leased land on the lick and sank the hollowed trunks of three sycamore trees ten feet into the ground. The three pipes served as wells. Using twenty-four kettles to evaporate the brine by burning almost five cords of wood, Brooks produced three bushels in a day. A decade later, the Ruffners, a family of inventive salt prospectors, made a new kind of drill by fitting an iron rod into a tapered wooden tube. A heavy wooden block was repeatedly raised and dropped on the rod, driving it into the ground, with the tube as a guide. But after seventeen feet, having reached solid rock, the rod would go no deeper. They then tried fitting a metal chisel to the rod, and the repeated pounding into the rock below slowly drilled a hole. This was considered a great innovation in drilling at the time, although the Sichuan Chinese had been doing the same thing since the twelfth century. In 1807, the Kanawha salt makers even invented a tube with a valve on the bottom to extract brine in the same way the Chinese had been doing it for 700 years.

By 1809, with these new inventions, Kanawha was experiencing a boom. Some fifty Kanawha salt producers had made their small riverfront the most important salt region in the United States after the Onondaga region. New holes were being drilled, and new boiling houses were being built. Most of the salt producers at Kanawha were short-term, small-scale operators who saw an opportunity to become wealthy in only a few years. In 1815, when fifty-five furnaces were operating, the largest producer had four. Most were single-furnace operations burning local coal.

The War of 1812 created the Kanawha saltworks' best years.

With no Liverpool salt and shortages everywhere, the number of furnaces increased in the three years of war from sixteen to fifty-two.

⌐

When you merely want to corn meat, you have nothing to do but to rub in salt plentifully and let it set in the cellar a day or two. The navel end of the brisket is one of the best pieces for corning.

A six pound piece of corned beef should boil three full hours. Put it in to boil when the water is cold. If you boil it in a small pot, it is well to change the water, when it has boiled an hour and a half; the fresh water should boil before the half-cooked meat is put in again.—*Lydia Maria Child,* The American Frugal Housewife, *1829*

THE GREAT OPPORTUNITY for Kanawha salt came with the postwar midwestern pork and beef industries. Because Kanawha salt could move inexpensively down the Ohio River, midwestern farmers produced tremendous quantities of pigs and cattle—especially pigs—and took them to the waterfront to be delivered to the river ports of Louisville and Cincinnati. There the meat was salt-cured and shipped throughout the settled parts of North America. Like the salt from the brine springs of Salies-de-Béarn, Kanawha salt was highly soluble and fast penetrating, ideally suited for curing meat.

The city of Cincinnati was built into a major commercial center with salt from Kanawha and pigs from Ohio, Kentucky, and Indiana. By the late 1830s, Cincinnati was packing almost one third of all western American hogs—more than 100,000 hogs per year. Other centers on the Ohio such as Louisville, Kentucky, and Madison, Indiana, also prospered.

Kanawha served this midwestern market with little competition. Unlike the French and the Spanish, English settlers and their American descendants tended to bring salt with them rather than find it where they went. In a market-driven society, a distant but efficient saltworks with good transportation seemed a more practical solution than a nearby but inefficient saltworks. As Americans moved west, they shipped salt from the East, just as the settlers in the East had shipped it from England.

Indiana, Illinois, and Kentucky saltworks, many of them adapted by the French from the saltworks of indigenous people, had weaker brine than Kanawha and used wood rather than coal.

A wood engraving of a slave working in a Virginia saltworks. The Granger Collection

This made Kanawha salt far cheaper to produce. In fact, even at outrageously inflated prices, Kanawha salt makers could still undersell their competitors, not only because of the density of their brine but because their workers were slaves.

Kanawha was in the state of Virginia, which had a huge slave-based tobacco industry that was slowly declining. The large tobacco plantations had more slaves than they could use, and the owners saw an economic opportunity in renting these people to Kanawha salt producers. According to the 1810 census, Kanawha county had 352 slaves, but by 1850, 3,140 slaves lived in the county, mostly assigned to saltworks.

By law the saltworks could only require slaves to work six days a week, but this law was seldom enforced. The best job for a slave at the saltworks was barrel maker. The salt was shipped in barrels, and slaves were expected to make seven barrels a day. Sometimes owners who rented their slaves would negotiate better terms such as only six barrels being required a day. The worst job in the saltworks, and one almost exclusively done by slaves, was coal mining. Slave owners would sometimes stipulate that their slaves were not to be used as coal miners, arguing that it was a misuse of their valuable property. Coal miners were often maimed or killed in cave-ins. The saltworks themselves were also dangerous, especially for slaves who were not trained in this industry. Boilers exploded, and sometimes workers would slip into near-boiling pots of brine. The owners sometimes sued the salt makers to be compensated for the loss or damage of their human property.

Plantation slaves did not want to be leased to saltworks, and they would sometimes escape while being taken west. The slaves knew that, like the salt, they were only a short journey to Ohio, which was a free state. Many slaves escaped by land or water, and salt makers would hire men to go to Ohio and bring them back. Once steamboats appeared, runaways increased, in part because of the transportation the boats afforded and in part because the boats hired free blacks who would encourage escape. Even the slaves who worked on the boats clearly lived a better life than the slaves in Kanawha.

In January 1835, Judge Lewis Summers complained, "There seems to be some restlessness among the slaves of the salt works and I thought more uneasiness in relation to that species of property than usual."

~

ROBERT FULTON DID not invent the steamboat, though he did build one of the first submarines, *The Nautilus,* which the French, British, and American governments all rejected. Fulton's enduring fame comes from a steamboat he launched in New York harbor in 1807 that sailed the Hudson. Numerous earlier steamboats had been built, and in 1790, John Fitch had established the first steamboat service, ferrying passengers between Philadelphia and Trenton. But such experiments were all commercial failures. Robert Fulton's Hudson River boat made money and demonstrated for the first time the commercial viability of steam-powered, flat-bottom, paddle-wheel-propelled riverboats.

These boats created Kanawha's first real competition. By the 1820s, steamboats for the first time made Liverpool salt accessible in the U.S. interior because they had enough power to carry a heavy salt load upriver against the strong currents of western waterways. The British liked to carry salt as ballast for their cotton trade with New Orleans, and they landed Liverpool, Turks, and Salt Cay salt there. The new steamboats carried foreign salt up the Mississippi River system, including the Ohio. The shallow-hulled steamboats could even navigate past the falls of the Ohio at Louisville, which had previously excluded Madison and Cincinnati from Mississippi traffic.

While this was happening, the Erie Canal opened, providing Syracuse with a western waterway. Before the Erie Canal, Onondaga salt had to be carried by pack mule to Lake Erie. The New York salt was preferred by many because, as with French bay salt, the slow solar evaporation process produced a desirable coarse grain. But because of slavery and ample nearby coalfields for fuel, Kanawha salt was cheaper.

No sooner was the Erie Canal opened than other canal projects were proposed. The first one was started in 1832: a 334-mile artificial waterway called the Trans-Ohio Canal, from the Ohio River to Cleveland on Lake Erie. Salt was the only bulk commodity transported on the Trans-Ohio Canal. By 1845, canals also connected Onondaga salt to the Wabash in Indiana.

The Erie Canal offered a refund on tolls to New York State salt producers if they used the canal to carry salt out of state. In trade on the Great Lakes, salt became ballast where empty ships had previously been weighted with sand. Sometimes they would carry the salt for free. By the 1840s, Syracuse, not Kanawha, became the leading supplier of salt in the Midwest.

⌒

IN THE 1840S, the Kanawha salt makers received another blow. The protective tariffs against imported salt designed to stimulate domestic salt production after the Revolution were angering westerners because they raised the price of salt. In their view, by taxing imported salt, the government was allowing domestic producers to overcharge. Kanawha, in particular, was the salt producer singled out for this accusation.

In 1840, Missouri senator Thomas Hart Benton delivered a speech in the U.S. Senate comparing Kanawha salt makers to the

British East India Company, a despised instrument of British colonialism, the economic system that Americans had fought two wars against.

> The tax on foreign salt, by tending to diminish its importation, and by throwing what was imported at its only seaport, New Orleans, into the hands of regraters, this tax was the parent and handmaiden of a monopoly of salt, which, for the extent of territory over which it operated, the number of people whom it oppressed, and the variety and enormity of its oppressions, had no parallel on earth, except among the Hindoos, in Eastern Asia, under the iron despotism of the British East India Company. . . . The American monopolizers operate by the moneyed power, and with the aid of banks. They borrow money and rent the salt wells to lie idle; they pay owners of the wells not to work them; they pay other owners not to open new wells. Thus, among us, they suppress the production, by preventing the manufacture of salt.—*Thomas Hart Benton, U.S. Senate, April 22, 1840*

Among the opponents of Senator Benton in the Senate was the delegation from Massachusetts, which desperately wanted to preserve the tariffs. But by the end of the decade the tariffs were removed, and not only Kanawha but also Cape Cod could no longer compete.

By 1849, when Henry David Thoreau visited the Cape, he was already writing about saltworks being broken up and sold for lumber. Those boards, used to build storage sheds, were still leaching salt crystals 100 years later. By then the Cape Cod salt industry was long vanished.

Kanawha survived. Soon the country would be divided into North and South, and it would become apparent that a southern saltworks was an important and all too rare asset.

CHAPTER SIXTEEN

The War Between the Salts

IN THE 1939 classic film of the Civil War, *Gone with the Wind*, Rhett Butler sneered at southern boasts of imminent victory, pointing out that not a single cannon was made in the entire South. But the lack of an arms industry was not the only strategic shortcoming of the South. It also did not make enough salt.

In 1858, the principal salt states of the South—Virginia, Kentucky, Florida, and Texas—produced 2,365,000 bushels of salt, while New York, Ohio, and Pennsylvania produced 12,000,000 bushels.

By 1860, the United States had become a huge salt consumer, Americans using far more salt per capita than Europeans. Numerous saltworks had sprung up in the North. Onondaga, the leading salt supplier, reached its peak production during the Civil War. The 200 acres of vats in 1829 had expanded to 6,000 acres by 1862, when it employed 3,000 workers and produced more than 9 million bushels of salt.

The United States as a whole was still dependent on foreign salt, but most of those imports went to the South. The imports from England and the British Caribbean were landed in the port of New Orleans. One quarter of all English salt entering the United States came through New Orleans. From 1857 to 1860, 350 tons of British salt were unloaded in New Orleans every day, ballast for the cotton trade.

As generals from George Washington to Napoléon discovered, war without salt is a desperate situation. In Napoléon's retreat from Russia, thousands died from minor wounds because the army lacked salt for disinfectants. Salt was needed not only for medicine and for the daily ration of a soldier's diet but also to maintain the horses of a cavalry, and the workhorses that hauled supplies and artillery, and herds of livestock to feed the men.

Salt was always on the ration list for the Confederate soldier. In 1864, a soldier received as a monthly allowance ten pounds of bacon, twenty-six pounds of coarse meal, seven pounds of flour or hard biscuit, three pounds of rice, one and a half pounds of salt— with vegetables in season. But the Confederate ration list was in reality a wish list that was only occasionally realized.

The Union army generally had ample supplies, and its rations included salt, salt pork, occasionally bacon, and both fresh and salted beef. Here too reality did not quite live up to the ration list. The salt beef, of which a Union soldier was issued one pound, four ounces per day, was greenish in color and unlovingly nick-named by the troops "salt-horse." John Billings, a Union veteran, writing about the rations after the war, mentioned numerous un-palatable recipes such as *ashcake,* which was cabbage stuffed with salted cornmeal and water and baked in ashes.

Four days after the war began on April 12, 1861, President Abraham Lincoln ordered a blockade of all southern ports. The blockade was enforced until the war ended in 1865. The North was able to put enormous resources into maintaining it. At its height in 1865, 471 ships with 2,455 guns were being used exclusively to enforce the blockade.

The blockade caused shortages and the accompanying high prices of speculators for not only salt but many basic foods. In 1864, potatoes cost $2.25 a bushel in the North and $25 a bushel in Richmond. Initially, the high prices posed more of a problem than scarcity. This recipe for salted beef appeared early in the war, when salt was still available but pork was unaffordable.

When Bacon gets too costly.

A gentleman who has tried the following recipe warmly recommends it: Cut the beef into pieces of the proper size for packing, sprinkle them with salt lightly, and let them be twenty four hours, after which shake off the salt and pack them in a barrel. In ten gallons of water, put four gallons salt, one pound saltpeter, half pound black pepper, half-pound allspice, and half gallon of sugar. Place the mixture in a vessel over a slow fire and bring to a boil. Then take it off and, when it has cooled pour it over the beef sufficient to cover it and fill the barrel. After the lapse of three or four days, turn the barrel upside down to be sure the beef is all covered by brine. If the beef is good, it will make it fit to set before a king. The beef will keep for a good long time. During the scarcity and exorbitant price of bacon our readers might try the recipe and test its virtue.—Albany Patriot, *Georgia, October 31, 1861*

When salt first started to become scarce in the Confederacy, owners of large plantations in coastal areas reverted to the Revolutionary War practice of sending their slaves to fill kettles with seawater for evaporation. But it soon became apparent that both the blockade and the war were to be far more serious than had been imagined. The small amount of salt produced from these kettles was not going to solve their problems.

At the outbreak of war, a 200-pound sack of Liverpool salt sold at the pier in New Orleans for fifty cents. After more than a year of the blockade, in the fall of 1862, six dollars a sack was considered a bargain in Richmond. By January 1863, the price in Savannah, a major port until the blockade, was twenty-five dollars for a sack.

The Union quickly realized that the salt shortage in the South was an important strategic advantage. General William Tecumseh Sherman, one of the visionaries of a modern warfare in which cities are smashed and civilians starved, wanted to deny the South salt. "Salt is eminently contraband, because of its use in curing meats, without which armies cannot be subsisted." he wrote in August 1862.

When the war finally ended, and Generals Ulysses S. Grant and Robert E. Lee sat down to talk, Lee said that his men had not eaten in two days and asked Grant for food. According to some observers, when the Union supply wagons were pulled into sight, the defeated soldiers of the famished Army of Northern Virginia let out a cheer.

⌒

IN 1861, THE western counties of Virginia organized into West Virginia, and Union general Jacob Dolson Cox marched in from Ohio up the valley of the Great Kanawha River. By July 1861, he controlled the entire valley, including the saltworks. It was one of the first major blows to the South. But in the fall of 1862, Confederate loyalists asked for volunteers to liberate the saltworks, and in a surprise attack a force of 5,000 Confederates drove the Union soldiers back to the Ohio River so quickly that they did not have time to destroy the saltworks before they retreated.

The Union learned a lesson from this: In the future, when they captured saltworks, they destroyed them. If the saltworks were brine wells, as in Kanawha—which Cox retook in November 1862, never to be retaken by the South—they broke the pumps and shoved the parts back down the wells. This contrasts with the Confederates, who when they took a saltworks celebrated having captured it and went into production.

A clerk in the Confederate War Department, blaming Confederate president Jefferson Davis's lack of resolve for the loss of Kanawha, wrote in his diary:

> The President may seem to be a good nation-maker in the eyes of distant statesmen, but he does not seem to be a good salt-maker for this nation. The works he has just relinquished to the enemy manufacturer: 7000 bushels of salt per day—two million and a half per year—an ample supply for the entire population of the Confederacy, is an object adequate to the maintenance of an army of 50,000 in that valley. Besides, the troops that are necessary for its occupation will soon be in quarters, and quite as expensive to the government as if in the valley. A Caesar, Napoleon, a Pitt, and a Washington, all great nation-makers, would have deemed this work worthy of their attention.

AS THE WAR went on, the Union army attacked saltworks wherever it found them, from Virginia to Texas. The Union navy attacked salt production all along the Confederate coast. At first, saltworks prospered on the Florida Gulf Coast because the area

was largely untouched by the war. By the fall of 1862, the Union
noticed the size and importance of Florida Gulf salt production
along the entire coast, but especially between Tampa, on the
mid–gulf coast, and Choctawhatchee Bay, on the western end
of the panhandle, near Alabama. The saltworks were usually hidden a few miles up inlets and were barely visible from the gulf.
Even if detected, the inlets were difficult for gunboats to navigate.

On September 8, 1862, the Union vessel *Kingfisher* approached
the saltworks at St. Joseph Bay on the panhandle under a flag of
truce and gave the Confederate salt workers two hours to abandon

An illustration from Harper's Weekly *of workers on the Florida Gulf
Coast fleeing with salt as the crew of the Union ship* Kingfisher *is about
to destroy the saltworks on September 15, 1862.* Collection of the New
York Historical Society

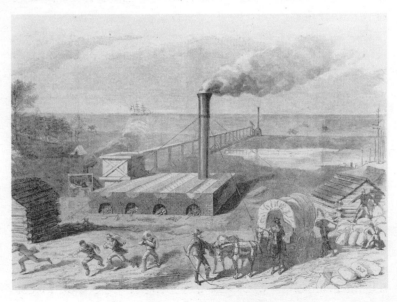

the site. Taking with them four cartloads of salt, the workers left. Three days later the Union navy destroyed the works.

On October 4, 1862, marines from the Union gunboat *Somerset,* moving farther down the coast, raided saltworks near Cedar Key on Suwannee Bay. After about twelve shells were fired, the salt workers raised a white flag. A landing crew met no resistance and destroyed several saltworks. But when they approached the one from which the white flag was flying, twenty-five men concealed in the rear of the building opened fire. Half the Union forces were wounded before reinforcements arrived from a nearby Union steamship. After driving the Confederate fighters into retreat, the Union landing party destroyed the boilers, an odd assortment of makeshift contraptions, and set the houses on fire. Some of the boilers and kettles were made of such thick iron that they had to fire howitzers at them to blow them apart. "The rebels here needed a lesson, and they have had it," said the commander of the *Somerset.*

The Union navy continued to attack saltworks on the Florida coast, burning houses, blowing apart equipment. By 1863, it had destroyed more than $6 million worth of saltworks back up at the panhandle in the St. Andrews Bay area. But saltworks, although easily destroyed, are just as easily and inexpensively rebuilt. In three months, many of the destroyed works were back in production.

Northern salt was smuggled into the South along with weapons, especially in Tennessee. Salt was also a common item of blockade runners. Liverpool salt was shipped to the Mexican port of Veracruz and from there to Brownsville, Texas, and into the Confederacy.

Mississippi governor John J. Pettus came up with an elaborate scheme to import 50,000 sacks of French salt in exchange for cotton brought to the shore of Lake Pontchartrain. The exchange was to be one bale of cotton for ten sacks of salt, arranged through the French and British consuls, to whose governments the Union blockade largely meant a loss of commerce. But though 500 bales of cotton were turned over to France, the French never delivered the salt.

WORKING CONDITIONS AT the improvised wartime saltworks of the Confederacy were even worse than those at Kanawha had been. At the saltworks that sprung up along the Tombigbee River, a few miles north of Mobile, Alabama, new wells were dug every day by people who had come from as far away as Georgia. During the war, southerners traveled hundreds of miles to the seacoast or a brine spring to make salt. The area along the Tombigbee, pro-tected by a Confederate fort in case the Union took Mobile, was marked by a vast traffic jam of carts and wagons filled not only with pots and boilers and other potential salt-making equipment but with poultry and other food—anything that could be traded for salt. Overseers drove the mule teams, and slaves followed be-hind on foot.

Some slaves chopped wood for fuel, the air shaking with hun-dreds of axes thudding into wood, while others dug fifteen-foot wells. In the beginning of the war, anyone could come and spend a few weeks making salt, but by 1862, leases regulated by the Al-abama legislature were required. By then the woods had been thinned out, and there was a fuel shortage. Shallow pans were de-signed for more efficient evaporation over two-foot-high furnaces with grates and iron doors. This equipment could make twenty to thirty-five bushels of salt a day, depending on the salinity of the water. Salt makers found that the deeper the well, the saltier the water, and began boring deeper into the bottoms of the original wells.

Slaves on long shifts kept the wells operating twenty-four hours every day. The saltworks were so close to each other that the area became a single undulating gnarled mass of slave labor. Land was set aside for a graveyard, which quickly filled as shivering slaves fell over from malaria or smallpox. Shoving and bumping against each other as they frantically labored to produce salt, slaves slipped and fell into boiling pans. Some died a quick death, but others died only after days of pain.

There were few white workers because most of the white men were drafted into the Confederate army. A handful of supervisors were draft rejects or wounded discharged veterans. As the war went on, more and more works were supervised by wounded veterans, usually amputees. In April 1862, when the first Confederate draft was declared, there were no exemptions for salt makers, but by August Jefferson Davis revised the conscription to exempt them. Making salt became a way to avoid military service. Deserters also drifted to the saltworks, hoping either to be safe in the swamps or to earn an exemption as a salt worker. In the last year of the war, the army searched wagons headed for the saltworks, looking for deserters among the exhausted slaves, the amputees, and the draft dodgers. By then, so many had deserted the Confederate army that there was even an organized deserter association in Virginia.

Refugees from attacked areas came to saltworks hoping to find a way to survive. Soon gamblers added themselves to the mix. Baptists and Methodists sent ministers to this increasingly iniquitous labor camp.

THE SHORTAGES IN the South presented opportunities to speculators. One way to earn a considerable fortune was to buy up a salt-producing area and control the local salt price. A single proprietor in Apalachicola controlled all of West Florida. To prevent such schemes, laws were passed in Georgia restricting coastline ownership.

Salt workers wanted to be paid in salt rather than money so that they too could profit from the inflated prices. Officials in the central government at Richmond, realizing the declining value of their money and the rising value of salt, stored large quantities of salt for possible barter arrangements.

A small packet of salt became a fashionable and much-valued gift. One such packet was a wedding present to George Edward Pickett, who later reached the most northerly point of any Confederate in combat when, on July 3, 1863, he led a ruinous charge up a sloping Pennsylvania field—the climax of the Battle of Gettysburg.

By 1862, Governor John G. Shorter of Alabama said, "The danger of a salt famine is now almost certain." From Mississippi, Governor Pettus wrote to Jefferson Davis that meat was being wasted and vanishing from diets because there was no salt to preserve the animals after slaughter.

A woman in South Carolina wrote:

It happened that my host at Radcliffe, just previous to the breaking out of hostilities, had ordered a boatload of salt, to use upon certain unsatisfactory lands [for fertilizer], and realizing that a blockaded coast would result in a salt famine, he hoarded his supply until the time of need should come. When it became known that Senator Hammand's salt supply was available, everyone from far and near came asking for it. It was like going down into Egypt for corn, and the precious crystals were distributed to all who came, according to the number in each family.

Family salt supplies were carefully hidden like a stash of jewels. Cheap salts cut with such substances as ash went on the market.

The *Tallahassee Sentinel* warns its readers of the folly of buying the dark and impure salt that is brought along the coast. It will not save meat but spoil it. We are informed that some of the salt makers, who are making for market, make an inferior article, for which they charge six and eight dollars a bushel. It were better to give twelve dollars or more per bushel and get a good article, than to buy that which is comparatively worthless at half the price. If our people will refuse to buy the inferior article it will soon induce salt makers to make a good salt. Pure salt is white, and that which is best for saving meat is large-grained. A word to the wise is sufficient.—Southern Confederacy, *Atlanta, August 28, 1862*

Rumors spread of possible salt substitutes. In 1862, there was a rumor of a substitute for curing bacon and beef. A newspaper in Alabama reported that pyroligneous acid, a vinegar made from hard wood, could preserve meat. A phenomenally popular British book of the period warned against it.

A very impure variety of pyroligneous acid, or vinegar made from the destructive distillation of wood, is sometimes used, on account of the highly preservative power of creosote which it contains, and also to impart the smoke-flavour; in which later object, however, the coarse flavour of tar is given, rather than that derived from the smoke from the combustion of wood.—*Isabella Beeton,* Beeton's Book of Household Management, *1861*

One southern publication suggested three ways to preserve fish without salt:

With oil: Put the fish in jars and pour over them salad oil until they are covered, then tie them up air tight. This is a rather expensive method in this country, but for fish that is afterward fried, it is excellent.

With acid: Dip them into or brush over them pyroligneous acid and then dry them by exposure to the air. This gives a smoky flavor, but if strong vinegar or pure acetic acid be used, no taste will be imparted. It may be applied by means of a painter's clean brush over a large surface. Fish and flesh so prepared will bear a voyage to the East Indies and back.

With sugar: Fish may be preserved in a dry state, and kept quite fresh by means of sugar alone, and even with a small quantity of it. Fish may be kept in that state for some days, so as to be good when boiled as if just caught. If dried and free from moldiness, there seems no limit to their preservation, and they are much better in this way than when salted. The sugar has no disagreeable taste. The process is particularly valuable in making what is called kippered salmon, and the fish preserved in this manner are far superior in quality and flavor to those which are salted or smoked. If desired, as much salt may be used to give the taste that may be required.—Southern Cultivator, *Augusta and Athens, Georgia, March–April 1863*

People tried curing beef with saltpeter and bacon with wood ash, neither of which worked very well. Newspapers were constantly revealing alternative techniques for curing, most of which were ineffective. Frequently these newspaper recipes, to inspire their suffering readers, made references to the salt shortages of the American Revolution. In 1861, a Richmond paper told the story of a Tory of Albemarle who had been refused salt because of his political sympathies. Still, his wife made good bacon with only one peck of salt and a large quantity of hickory ashes.

> In applying the ashes, it is well to have a bucket of molasses, and apply a portion with a white-washing brush to each joint. When well smeared, rub on the ashes, which will thus adhere firmly and make an impenetrable cement.—Daily Richmond Examiner, *November 23, 1861*

New ideas for salt conservation were a constant topic between neighbors. Those who lived by coasts would cook their starch— rice, hominy, or grits—in seawater, which would provide the only salt in a meal. When eating salted meat, people would carefully brush off every loose salt crystal for reuse. The brine in troughs and barrels used for pickling was afterward boiled down and made into salt to use again. The earth around smokehouses, made salty from years of drippings, was dug up and placed in hoppers designed for leaching ashes in soap making. This technique yielded a brine that could be evaporated, leaving a dull, darkish bed of salt crystals.

> Coal consists mainly of the carbon in wood, which in burning forms a very dry heat. Most of our readers are familiar with the usual process of barbecuing large pieces of meat over coals. If such meat were too high above the coal-fire to roast, it would soon dry. When dry, a very little salt and smoking will keep it

indefinitely. Like cured bacon, it should be packed in tight casks, and kept in a dry room.

After one kills his hogs, if he is short of salt, let him get the water out of the meat by drying it over burning coals as soon as possible, first rubbing it with a little salt. Shade trees around a meat house are injurious by creating dampness. Dry meat with a coal-fire after it is smoked. You may dislike to have meat so dry as is suggested, but your own observation will tell you that the dryest hams generally keep the best. Certainly sweet, dry bacon is far better than moist, tainted bacon, and our aim is simply to show how meat may be cured and long kept with a trifle of salt, when war has rendered the latter scarce and expensive.—*Dr. F.P. Porcher of the Confederate army,* Economy in the Use of Salt, *1863*

HOW TO MANUFACTURE SALT FOR HOME USE

Take a towel, or any piece of cloth—say, two yards long—sew the two ends together, hang it on a roller, and let one end revolve in a tub or basin of salt water; the sun and air will act on the cloth, and evaporate the water rapidly. It must be revolved several times throughout the day, so that the cloth is well saturated. When the solution is evaporated to near the bottom, dip from the concentrated brine and pour it in a large flat dish or plate; let it remain in the sun until the salt is formed; taking it in every night, and placing a cover on it. This is accomplished by capillary attraction, and can be manufactured for $1 per sack, on a large scale. Each gallon of salt water will produce two and a half ounces of salt when evaporated.

p.s. To make salt requires a little patience, as it is of slow formation.—*John Commins, Charleston tannery,* Charleston Mercury, *June 11, 1862*

⌒

JUST BEFORE THE war began, French geologist M. J. Raymond Thomassy wrote that Louisiana, with its sugar and cotton, needed only to add salt production to its economy to become truly wealthy. He warned:

> Just as this element of future prosperity, this vital food, almost as necessary to their economic independence as gunpowder has been to the national independence, is furnished them exclusively by strangers, and is found in hands, which, in spite of all the dreams of perpetual peace, could easily some day become those of an enemy, and be made into an instrument, if not of domination, at least of famine and internal trouble.—Géologie Pratique de la Louisiane, 1860

Thomassy had a theory, dismissed by most people in Louisiana at the time, that certain areas in the southern part of the state— in particular, a swampy area once known as Petite Anse, meaning "little harbor"—were sitting on beds of rock salt. Petite Anse was covered with ferns and long-rooted trees dripping in moss and broad-leafed growths so thick, only a skilled local would be able to see it was an island, a raised area of 2,200 acres, surrounded by dark waterways connecting bayous that ran into the Mississippi and the nearby Gulf of Mexico at a harbor called Vermilion Bay.

When John Hayes settled in Petite Anse in 1791, salt had been made there for a very long time. In a slight variation on the Salies-de-Béarn and Lüneburg pig-in-the-marsh story, Hayes was supposedly hunting deer, not boar, and it was he, not his prey, who discovered the brine, when he stopped to drink some water.

Soon after Hayes found this natural brine spring in Petite

Anse, a man named Jesse McCaul bought nineteen acres there and began making salt. He dug several wells, and at a depth of three to six inches, he found pottery fragments. Later it would be discovered that these fragments were spread over a five-acre area, the site of a prehistoric saltworks in the manner of the early Romans, where brine was evaporated in pottery and then the pots were broken. Piles of ancient shards are occasionally still found on the island. Archaeologists believe these saltworks are 1,000 years old. But recently a mound was found on the island in a place called Banana Bayou that was carbon-dated to about 2500 B.C., which would make it one of the oldest man-made structures ever found in the United States.

Neither Hayes nor McCaul did much with the salt of Petite Anse, the island in the swamp with invisible shores. McCaul tried, digging several wells, but failing to make a profit, he abandoned the project. When the War of 1812 drove salt prices up, sending entrepreneurs looking for brine, a man named John Marsh turned Petite Anse into a profitable saltworks.

In 1841, New Orleans was the third largest city in the United States, a leading port, and an obvious destination for Edmund McIlhenny of Maryland, seeking his fortune in banking. In Creole New Orleans, dominated by the descendants of French and Spanish settlers, people like McIlhenny were immigrants called "Americans." Working his way up from bookkeeper in this cosmopolitan city, already famous for its local cuisine and foreign restaurants, by 1857 McIlhenny had five banks in Louisiana and was a wealthy man enjoying the luxuries of his adopted metropolis. He befriended a Baton Rouge judge, Daniel Dudley Avery, who was only five years older. Avery had married a Marsh and come into possession of Petite Anse, which he used as a sugar plantation.

In 1859, to the shock of some, the middle-aged McIlhenny

Edmund McIlhenny. McIlhenny Company, Avery Island, Louisiana

married his friend's young daughter, Mary Eliza Avery. Wishing to escape the war and continue their sumptuous Louisiana lives, both the McIlhenny and Avery families moved to the sheltering dark bayous of Petite Anse. There they might have lived out the war quietly, had it not been for the discovery that Thomassy, the French geologist, had been right.

On May 4, 1862, a slave at the bottom of a sixteen-foot hole, while attempting to clean and deepen a brine well, said he had hit a log that he could not remove. Upon investigation, it was found that the obstruction was solid salt. Petite Anse was sitting on a bed of solid, remarkably pure salt, estimated to be about forty feet deep, with 7 million tons of salt. Generations later, it would be realized that this estimate had been far too modest.

The discovery, although it was exactly what Thomassy had predicted, came as a great surprise. The salt was especially valuable because it was much purer and drier than most rock salt. It was extremely hard and had to be blasted with dynamite, which yielded great jagged chunks of white crystal. To transport the salt, the two families built a two-mile-long causeway across the bayou and swampland to the town of New Iberia.

Suddenly the genteel McIlhennys and Averys found themselves sitting on a strategic war target. They began producing salt for the South. Judge Avery was flooded with offers for contracts. Governor Pettus wrote fellow Mississippian Jefferson Davis that there was at Petite Anse "salt for all the Confederacy." Newspapers ran reports of similar salt finds, but most were false rumors.

Union forces made several attempts to take Petite Anse, and the families fled to Texas. In January 1863, the Union sent a steamer and two gunboats to Vermilion Bay, two miles from the island saltworks. That night, the wind shifted to the north and drove the water from the bayou, and by morning the two gunboats were aground in mud, where they remained stuck for the next twenty days. But on April 17, 1863, a Union colonel took his troops south of New Iberia and attacked the saltworks, destroying eighteen buildings with their steam engines, boiling and mining equipment, as well as 600 barrels of urgently needed salt that was about to be shipped throughout the Confederacy.

The Union troops were surprised at how easily they took this major saltworks and interpreted the inability of the Confederates to defend this strategic point as a sign that the South was crumbling.

But some of the bloodiest battles of the war were yet to come. With the help of liberated slaves, the Union continued to cripple the southern war effort by attacking saltworks—Darien,

Georgia, and Back Bay, Virginia, in September, and Bear Inlet, North Carolina, on Christmas Day. The following year, it destroyed saltworks at Goose Creek, Florida; Masonborough Inlet, North Carolina; Cane Patch, South Carolina; Tampa and Rocky Point, on Tampa Bay; and Salt House Point, Alabama. On December 10, the day Sherman completed his march of destruction through Georgia, troops under George Stoneman marched from Knoxville, Tennessee, with the objective of destroying saltworks and supply depots in eastern Tennessee and southwestern Virginia. On December 20, Stoneman's troops destroyed the saltworks at Saltville, Virginia. On February 1, 1865, one last time, the Union navy destroyed the saltworks in St. Andrews Bay on the Florida panhandle.

Even civil wars produce occasional acts of kindness. Eighteen days after the final attack on St. Andrews Bay, General Oliver Otis Howard, having taken Columbia, South Carolina, ordered that before the storehouses were destroyed, the Columbia hospital was to be furnished with as much salt as it needed and that more salt be saved for the poor who had been burned out of their homes.

TO KEEP MEAT FROM SPOILING IN SUMMER
Eat it early in the Spring.—Confederate States Almanac, *Macon, Georgia, 1865*

CHAPTER SEVENTEEN

∽

Red Salt

AFTER THE WAR ended, with more than 1 million Americans dead, Dudley Avery returned to Petite Anse. He had fought for the Confederacy and survived such battles as Shiloh, Tennessee, where 1,723 Confederates and 1,754 Union soldiers were killed in a standoff. Not only did Judge Avery's son survive, but the Judge's finances remained sound enough that he could buy the remaining third of the 1,600-acre island, and Petite Anse became Avery Island, a single-family property for the first time in history.

Edmund McIlhenny and his wife returned from Texas, where he had offered his business skills in the service of the Confederate army's commissary and paymaster's office. Before he had fled, McIlhenny had earned a considerable fortune from salt. But he had accepted payment in Confederate bills. Knowing that the salt prices on which he had earned this mountain of useless money were not going to come again, he went to New Orleans in search of new business opportunities.

Postwar New Orleans offered few opportunities for an out-of-work banker. At this critical moment, McIlhenny's story becomes uncertain because he left no record of the events. All that remains are the recollec-

Capsicum frutescens tions of various relatives who had been told parts of the story. Apparently, a man came up to McIlhenny on the street. In one version he was an old veteran of the 1846 Mexican-American War. In a more probable version, he was a Confederate veteran who had fled to Mexico to avoid being taken by the Union army. This man, whose name was Gleason, was very excited about a certain Mexican seasoning, small red chili peppers.

In 1866, unsuccessful in resurrecting his business career, McIlhenny returned to Avery Island and resolved to become a gentlemen farmer, experimenting with hot peppers.

THE BURN OF a pepper comes from a substance called capsaicin, which is a natural poison designed to protect the plant by making it inedible. But Mexicans, Caribbeans, and a great number of other people have not been deterred. Capsaicin develops in sunlight and certain soils. With peppers, as with wine grapes, the place where they are grown makes all the difference. The peppers that Edmund McIlhenny brought home, subsequently labeled *Capsicum frutescens,* when grown on the fertile soil on the edges of Avery Island, were extremely hot.

The idea of a pepper sauce was not new to southern Louisiana. The Cajuns, French refugees who fled Nova Scotia after it fell to the British in the eighteenth century, had settled along the bayous in the Avery Island area, and they, like the Creoles of New Orleans, had learned to use hot peppers brought by the Caribbean and Mexican people who came through the port. Before the Civil War, New Orleans cooks dried hot peppers and marinated them in sherry and vinegar. Red pepper and salt were already a common seasoning blend in Cajun cooking.

McIlhenny's wife, Mary Eliza Avery, left a handwritten collection of recipes. Since she used her maiden name on the front page, the collection can be dated before 1859, the year of her marriage. In this collection of Cajun and southern Louisiana recipes, numerous dishes call for "red pepper and salt."

SHRIMP GUMBO

Take a chicken and cut it up as for a fricassee. Put in your soup pot a spoon full of lard, when hot stir in two table spoons full of flour until it becomes a lite brown color; chop fine a large onion, and throw into the flour and lard with the chicken, stirring it until the chicken becomes slightly cooked. Add boiling water as [appropriate?] for a soup stirring well. Put in red pepper and salt to your taste, with a bunch of parsley and thyme (this preparation can be made of shrimp as well as chicken). Take 4 or 5 pints of shrimps and pour boiling water on them when they first come from market. Take the meat and the roe from the shells. Put the heads and shells into a stirr-pan, covering them with boiling water. Mash them so as to extract the juice—strain it and add the liquid to the soup. About 15 minutes before sending to table throw in the Shrimps. When ready to serve the soup, stir in a large tablespoonful of fresh Fillet and turn immediately into the tureen.—*Mary Eliza Avery*

McIlhenny started his pepper sauce experiments with a variation on a sauerkraut recipe, using salt to ferment and extract juices from fresh crushed peppers. He quickly learned that he had to use the ripest peppers, picking each of the fruits of his annual plant at its optimum moment, when it was the brightest red. Stirring half a cup of his own Avery Island salt into each gallon, he aged the mixture, trying pickling jars and then pork barrels. He covered the lids with salt, which, when mixed with the juices of the fermenting peppers, sealed the barrel with a hard crust, by chance the same way the Chinese had been aging bean mash for soy sauce for thousands of years.

Up to this point, it was an all-Avery-Island product, made with the island's salt and peppers. In much of the South, the Caribbean, and Mexico, this would constitute a hot sauce. But the New Orleans tradition called for vinegar. McIlhenny strained the mash after it had aged for one month, and mixed it with French white wine vinegar. Then he put it in small cologne bottles, which he sealed with green wax. Each bottle came with a little sprinkler attachment that could be placed in the opening after the seal was broken.

McIlhenny had a shed on Avery Island that he called his laboratory, a place with a sweet pungent smell that tickled the nostrils and made passersby want to sneeze. He would let his children leave school early to help him in the laboratory.

In 1869, he produced 658 bottles and sold them for the handsome price of one dollar each wholesale in New Orleans and along the Gulf. People used the sauce as a seasoning in recipes that called for red pepper and salt. In 1870, he obtained a patent and named the concoction Petite Anse Sauce. His family was appalled that he was commercializing the historic family name for his eccentric project. So he settled for Tabasco sauce, after the Mexican state on the Gulf known for hot peppers, possibly even the area where the mysterious Gleason had obtained the original peppers.

April 28, 1883, Harper's Weekly *drawing of salt mining under
Avery Island.*

These were not lucrative years on Avery Island. An attempt
to return to salt mining was a failure. By 1890, when Edmund
McIlhenny died at age seventy-five, he had built a modest family business with Tabasco sauce, though nothing compared to the
fortune he had earned in useless currency from two years of
wartime salt.

⌒

AFTER THE CIVIL War, while pepper sauce was looking more
lucrative than salt in Louisiana, the American West offered dramatic opportunities. The West was rich in precious minerals and
in salt, which was still used to purify ore, especially silver.

The most spectacular salt strike in North America was found
in a shrinking glacial lake in Utah. In the eighteenth century, the

Spanish, while searching Utah for gold and silver, had been told of a huge salt lake. But they never saw it. The first record of anyone of European origin seeing the Great Salt Lake was in 1824 by James Bridger, a hunter, trapper, and explorer who was the prototype of the legendary "mountain man."

In Carthage, Illinois, in 1846, an angry mob assassinated Joseph Smith, the leader of a new religious group known as the Mormons. Brigham Young, who took Smith's place, wanted to find a new land where Mormons could set up their own community away from the scrutiny of other Americans. In search of a place with natural resources, so that his isolated group could have economic self-sufficiency, he chose this Great Salt Lake in the middle of a desert that at the time belonged to Mexico. The lake had no outlet and contained highly concentrated brine. Next to it was one of the largest sebkhas ever found—a flat, thick, 100-mile-long layer of salt, which became a mainstay of the Mormon economy.

Other salt beds were found in the West, but none so large or with as pure a concentration of sodium chloride as the Great Salt Lake area, the remains of a far larger 20,000-square-mile prehistoric lake geologists call Lake Bonneville.

But the real need for salt lay farther west in Nevada and California, where silver was found. Relatively close to these silver strikes was one of the oldest saltworks in the American West.

The southern end of San Francisco Bay is an insalubrious marshland with ideal conditions for salt making. Not only does it have more sun and less rainfall than San Francisco and the north bay, but it has wind to help with evaporation. The intensely hot air from central California comes over the mountains, and the temperature difference sucks in the cool sea breeze.

This is why centuries and perhaps millennia before the California and Nevada silver strikes, a people called the Ohlone made annual pilgrimages to this area for salt making. At the water's edge, the brine slowly evaporated in the sun and wind and left a thick layer of salt crystals. They had only to scrape it. The first European to notice the local salt making was a Spanish priest, José Danti, who explored the eastern side of the bay in 1795. In the southern end of the east bay, he found marshes with thick layers of salt, and "the natives," he reported, told him that it provided salt for much of the area.

The Spanish were content to let the Ohlone produce salt. They only wanted a share—a very large one—of the profits. To this end, they forced the Ohlone to turn all their salt over to the Spanish missionaries who controlled distribution. The only technology added by the Spanish was to drive stakes into the ground at the water's edge to offer additional evaporation surfaces.

In 1827, Jedidiah Smith, one of the earliest U.S. citizens to settle in California, arrived in San Francisco Bay and noted that "from the Southeast extremity of the bay extends south a considerable salt marsh from which great quantities of salt are annually collected and the quantity might perhaps be much increased. It belongs to Mission San Jose."

After California became a state in 1850, a San Francisco dockworker named John Johnson became interested in this salt area. At age thirty-two, his life story was already a popular legend. Supposedly, he had lost both parents in a fire from which he was saved as a baby in Hamburg, Germany. He went to sea at thirteen and was one of two hands that survived on a sinking ship by perching on top of the highest two masts for twelve hours. He was said to have been a sealer, a whaler, and a slaver—a ruthless adventurer who would try anything to make money. When he learned about the southeast part of the bay, he decided to try salt.

At first, Johnson was able to charge extremely high prices

and make tremendous profits. But this was the time of the gold rush, and adventurers from all over the world were coming to the Bay area looking for quick profits. Many followed Johnson to the south bay. Soon abundance caused the price to crash.

San Francisco Bay salt was considered of low quality, and it did not compete easily with Liverpool salt, which came as ballast on the British ships that bought California wheat. Coarse salt was also regularly imported from China, Hawaii, and numerous places in South America.

But in 1859, something happened that drove salt prices back up. In western Nevada near the California border, a three-and-a-half-mile stretch of the Sierra Nevada mountains was found to hold the richest vein of silver ever discovered in the United States. It was called the Comstock lode, named after an early investor who had sold out before the extent of the vein was known. The silver ore was being separated by a technique similar to the sixteenth-century Mexican patio process, and it required mountains of salt.

By 1868, only nine years after the discovery of the Comstock lode, eighteen salt companies were operating in the south bay. To keep the profit margin high, the salt was mostly harvested by Chinese laborers, the cheapest source of labor in California at the time. The salt workers wore wide wooden sandals to avoid sinking in the thick layer of white crystals.

After a few years of scraping, the area was running out of naturally evaporated salt, and the salt producers started building successive artificial ponds, pumping the water from one pond to the next by the power of windmills.

Fortunes were being made on the silver in Nevada and on the salt in California. Then, in 1863, a man named Otto Esche came

A nineteenth-century photograph of a windmill pumping brine into a saltworks in south San Francisco Bay. The Bancroft Library, University of California, Berkeley

up with a scheme to make money on the link between them. The salt was shipped inland and into the Sierra Nevadas in horse-drawn carts. Esche went to Mongolia, even today one of the more remote corners of the earth, and bought thirty-two Bactrian camels. Esche apparently knew something of camels since he chose the more docile two-humped camel rather than the notoriously temperamental dromedary of the Middle East. Bactrian camels, since before Marco Polo's time, had been carrying goods, including salt, across the brown, wide, Mongolian desert.

The first discouraging surprise was that only fifteen camels survived the cross-Pacific voyage to California. The survivors arrived in such bad condition that it took Esche months to nurse them back to good health. They carried salt across the mountains, but the strange, furry, long-legged creatures were not well received in Nevada.

The silver miners can be added to a long list of novices who have found that camels, even the better-tempered Bactrians, can be disagreeable. They bite, spit, and kick. The miners hated them, as did their horses and mules, who became hysterical at the sight of them. This reaction by the other animals made the camels a public nuisance. A few would lope into town, and suddenly the street was alive with neighing, braying, and kicking. Virginia City, Nevada, passed an ordinance outlawing camels on the town streets except between midnight and dawn, when, presumably, the other animals were in stables resting. Eventually, to the relief of the miners, Esche gave up on the camels and released them in the Nevada desert to thrive on their own. Since no camel colony has ever been discovered there, it is assumed they all died, probably a slow, pitiful death.

⌒

IN THE SPRING, seawater was pumped into the ponds of the south bay. Through the summer, the brine would be moved; by late summer, it was dense enough to crystallize. The brine turned pink and then a dark brick color. Today, when people fly into San Francisco, they sometimes gaze out the window and wonder about the pink-and-brown geometric ponds at the end of the bay.

The color is a common phenomenon that had previously been observed in Europe, in the Dead Sea, and in China, among many other places that made sea salt. Both Strabo and Pliny wrote about this curious color in brine that later disappeared after crystallization. Strabo, who pondered the color of parts of the Red Sea, thought it was caused by either heat or a reflection.

In *Salt and Fishery, Discourse Thereof,* Collins had mentioned the phenomenon, attributing the color to red sand. The red color was generally thought to be an impurity that could cause spoilage, and it was believed that it might turn the fish or meat red and then the food would soon spoil. In 1677, Anton van Leeuwenhoek, the Dutch naturalist who made numerous discoveries with a crude microscope, concluded that the red color was caused by microorganisms in the brine.

Whatever the cause, the simple observable fact, as Denis Diderot pointed out in his eighteenth-century encyclopedia, is that "you know the salt is forming when the water turns red."

Charles Darwin observed the phenomenon in Patagonia:

> Parts of the lake seen from a short distance appeared of reddish color, and this perhaps was owing to some ifusorial animalcula. The mud in many places was thrown up by numbers of some kind of worm, or annelidous animal. How surprising it is that any creatures should be able to exist in brine.

In 1906, E. C. Teodoresco identified a one-celled plant called dunaliella, which most observers concluded must actually be two species because the brine initially developed a green scum and only later, when more dense, turned red. Were there both green and red dunaliella? Darwin wrote of the complex ecology of sea saltworks where single-celled algae lived in brine and turned it green, but at a denser level, tiny shrimp and worms turned it red, and these reddish animals attracted flamingos, which turned pink from eating them. In fact, Darwin had figured out the entire mystery in the nineteenth century, but few listened to him until well into the twentieth century.

The San Francisco Bay salt makers of the silver rush days believed the dark red color came from insects in the brine. Only in modern times has it been understood that dunaliella is green, but once the brine reaches a certain level of salinity, it turns red. In addition, tiny, barely visible shrimp, brine shrimp, live in the brine at this density. And there are also salt-loving bacteria of reddish hue that are attracted to brine. Not only does the red color signal that the brine is ready, it intensifies the solar heat and hastens evaporation, helping the salt to turn to crystals and fall out of the reddish water. Today, the saltworks of San Francisco Bay sell their reddish little creatures to other saltworks that wish to improve their evaporation process.

Just as Diderot had observed but could not explain, when the brine reaches the density that attracts these shrimp, algae, and bacteria, it means that the brine is at a density close to the point of crystallization. The process of making salt, though practiced since ancient times, was beginning to be understood.

PART THREE

Sodium's Perfect Marriage

It is an old remark, that all arts and sciences have a mutual dependence upon each other. . . . Thus men, very different in genius and pursuits, become mutually subservient to each other; and a very useful kind of commerce is established by which the old arts are improved, and new ones daily invented.
—William Brownrigg,
The Art of Making
Common Salt, *London, 1748*

CHAPTER EIGHTEEN

The Odium of Sodium

Edmund Clerihew Bentley, a British author of crime novels who lived from 1875 to 1956, wrote these lines, it is said, while in a chemistry class:

Sir Humphry Davy
Abominated gravy.
He lived in odium
Of having discovered sodium.

This was the first of a verse type known as a clerihew, which is a pseudo-biographical verse of two rhymed couplets in which the subject's name makes one of the rhymes. It became a genre of humorous poetry, although not many people can recite another example of a clerihew.

Sir Humphry Davy was also an Englishman, born in 1778, and a largely self-taught chemist. When he was a twenty-year-old apprentice pharmacist in Cornwall, the Pneumatic Institution of Bristol offered him a job researching medical uses of gases, which may have been a twenty-year-old's dream job. There is little evidence of his feelings about gravy, but he was known to have a great fondness for nitrous oxide, laughing gas, which he experimented with at length and found to be not only an enjoyable recreational drug but a cure for hangovers. Notable friends, including the poets Robert Southey and Samuel Taylor Coleridge, shared in the experiments. But Davy learned that some gases are better than others, and he almost died from his experiments with carbon monoxide.

Davy's work in Bristol came under attack by conservative politi-

Young Humphry Davy merrily mans the bellows in a laughing-gas experiment being presented by his predecessor at the Royal Institute, Benjamin Thomas, in a caricature by James Gellray. The Chemical Heritage Foundation, Philadelphia

cians, including the famous Irish MP Edmund Burke, who accused the gas experiments of promoting not only atheism but the French Revolution.

But within a few years, his other experiments with electrolysis, passing electricity through chemical compounds to break them down, earned him enduring fame. Davy's chemistry lectures at the Royal Institute became so noted for the brilliance of his delivery that the talks were regarded as fashionable cultural events. Then, in 1812, to the disappointment of fans, he married a wealthy widow and gave up lecturing to spend his time touring Europe.

Davy, a brilliant scientist, had a flair not only for performance and for living well, but also for self-promotion. He managed to garner credit for a phenomenal number of the scientific breakthroughs of his day. Through electrolysis, he was able to isolate for the first time a number of elements, including, in 1807, sodium, the seventh most common element on earth. This discovery was the first important step toward at last understanding the true nature of salt.

THAT DIFFERENT TYPES of salt existed and were suited for different purposes was a very old idea. The ancient Egyptians knew the difference between sodium chloride and natron. But they didn't understand their composition or how to make them. Saltpeter, which can be sodium nitrate or potassium nitrate, was well known by the medieval Chinese, who used it for gunpowder. After Europeans learned about gunpowder, the market for potassium nitrate seemed limitless. But little was known about its properties.

As early as the sixteenth century, nitrates were used in cured meats to make them a reddish color, that was thought to be more in keeping with the natural color of meat. In fifteenth-century Poland, game was preserved in nitrate simply by gutting the animal and rubbing the cavity with a blend of salt and gunpowder, which was potassium nitrate. It took centuries of use before anyone understood how potassium nitrate and its cheaper cousin, sodium nitrate, which is often called Chile saltpeter, are broken down by bacteria during the curing of meat. The nitrate turns to nitrite, which reacts with a protein in the meat called myoglobin, producing a pinkish color. The reaction also produces minuscule amounts of something called nitrosamines, which may be cancer-causing. Today, the amount of nitrates is limited by law to what seems to have been deemed an acceptable risk for the oddly unquestioned goal of making ham reddish.

For centuries, different types of salts were recognized by taste. The Great Salt Lake was clearly a concentration of sodium chloride because it had a pleasant salty taste, whereas the "bitter nauseous" taste of the Dead Sea indicated magnesium chloride. The long practiced principle of evaporating brine was that when brine becomes supersaturated—when it is at least 26 percent salt, which is considerably more than the 2.5 or 3 percent salt of seawater—sodium chloride crystalizes and falls out, or precipitates, from the liquid. But slowly it was discovered that after the sodium chloride, the salt of primary interest, precipitates, a variety of other salts crystalize at even denser saturation.

In 1678, Dr. Thomas Rastel of Droitwich wrote:

Besides the white salt above spoken of we have another sort called clod salt, which adheres to the bottom of the vats and which after the white salt is laded out, is digged up with a steel picker. This is the strongest salt I have seen and is most used for salting bacon, and neat's [ox] toungues: it makes the bacon redder than other salt, and makes the fat meat firm.

⌒

THE WORD *CHEMISTRY* was first used in the early 1600s, although the science was not considered an independent field of research until the end of the century. One of the accomplishments of early chemists had been to identify some of the salts that precipitated out of brine. But despite this work, it seems that very few people in the seventeenth century had any idea what a salt was.

A 1636 book by Bernard Palissy, with the dreamy title *How to Become Rich and the True Way in Which Every Man in France Could Grow and Multiply Their Treasury and Possessions*, states that "sugar is a salt." In listing all the "various salts," Palissy includes "grape salt, which gives taste and flavor to wine." It is not surprising that he concluded that it was impossible to list all the salts. In John Evelyn's 1699 discourse on salads, he states that sugar is sometimes referred to as "Indian salt."

Apparently, there was little definition of salt other than as something made of white crystals. This began to change in the early seventeenth century, when Johann Rudolf Glauber, a German chemist, took a cure in a spring near Vienna and extracted from the water a salt that he called *sal mirabile*. The salt was hydrated sodium sulphate, though Glauber could not have put it that way, because Davy had not yet discovered sodium. Glauber sold his discovery as a secret cure, a mineral bath of allegedly wondrous health benefits. It became so famous that today, though it is more used in metallurgy, textiles, and other industries than as a bath salt, it is still commonly known as Glauber's salt.

Enough of an entrepreneur to keep his formula secret, Glauber was also enough of a scientist to reveal, after his fortune was made, that when sulphuric acid was applied to common salt, producing hydrochloric acid, a process that had already been well known for centuries, the residue, that had always been thrown out, was Glauber's salt.

Later in the same century, Nehemiah Grew, a British plant physiologist who is credited with being the first human ever to witness and document plants having sex, studied the celebrated health spring water of Epsom in Surrey, England. He isolated a salt, magnesium sulphate, ever after known as Epsom salt. Epsom salt is now used not only medicinally but in the textile industry, for explosives, in match heads, and in fireproofing.

But Nehemiah Grew was even less forthcoming than Glauber about his discovery. Only after years of speculation was it discovered by chemist Caspar Neuman in 1715 that Epsom salt could be made by applying sulphuric acid to the mother liquor.

Mother liquor is the dark blood-red water that remains after common salt precipitates out of brine. An eighteenth-century London chemist named John Brown discovered that Epsom salt could be boiled out of the mother liquor without sulphuric acid. Brown also found another salt in the liquid. The study of this third salt, now known to be magnesium chloride, unleashed a chain of discoveries, including Davy's 1808 announcement that he had found a new element, magnesium. In 1828, Antoine Bussy isolated workable quantities of the metal, and an industry was born. Magnesium is used to prevent corrosion of steel and in explosives, lightbulbs, and lightweight metal alloys.

At the time of Neuman's early-eighteenth-century experiments with mother liquor, the liquid was called bittern, and salt makers usually threw it away or fed it to animals or even poor people as a cheap source of salt. The Dutch found that it worked well for washing windows. Despite pleas from scientists, most saltworks continued to throw out their leftover bittern.

Then, in 1792, sodium carbonate, soda, was made from mother liquor. Soda found in nature had been used since ancient times in early industries such as glassmaking. Natron is a form of soda. In fact, Davy named sodium after soda because it was one of the element's best known compounds. The manufacture of artificial soda started numerous industries. Sodium hydrogen carbonate, bicarbonate of soda, is used in food as well as for glassmaking and textiles. Sodium carbonate is used in making paper, plastics, detergents, and the artificial fabric rayon.

By the time of the Civil War, commercially made soda was common, and soda fountains had already become widespread in America. A popular American women's magazine gave a recipe for making carbonated drinks.

> Put into a tumbler lemon, raspberry, strawberry, pineapple or any other acid syrup, sufficient in quantity to flavor the beverage very highly. Then pour in very *cold ice-water* till the glass is half full. Add half a teaspoonful of bicarbonate of soda (to be obtained at the druggist's) and stir it well in with a teaspoon. It will foam up immediately, and must be drank during the effervescence.
>
> By keeping the syrup and the carbonate of soda in the house, and mixing them as above with ice-water you can at any time have a glass of this very pleasant drink; precisely similar to that which you get at the shops. The cost will be infinitely less.—Godey's Lady's Book, *1860*

For many centuries there had been a great confusion between potash and soda. The name potash is derived from the process used for making potassium carbonate, cooking down water and wood ash in earthen pots. Like soda, potash had many industrial applications long before it was chemically understood. Among other things, it was used in making glass and soap.

Before sodium carbonate in the form of baking soda was manufactured, potash was used in baking. Amelia Simmon's cookbook, originally published in Hartford and then Albany in 1796, is considered the first American cookbook not only because it was published after the Revolution, but because it was written by an American, for Americans. Simmons used enormous quantities for apparently huge cakes. One recipe, for "Independence cake," called for twenty pounds of flour, fifteen pounds of sugar, ten pounds of butter, and twenty-four eggs. Many of her baking recipes called for "pearl-ash," which was potash, as a rising agent.

HONEY CAKE

Six pound flour, 2 pound honey, 1 pound sugar, 2 ounces cinnamon, 1 ounce ginger, a little orange peel, 2 tea spoons pearl-ash, 6 eggs; dissolve the pearl-ash in milk, put the whole together, moisten with milk if necessary, bake 20 minutes.—*Amelia Simmons,* American Cookery, *1796*

In 1807, when the potash industry was already many centuries old, Davy connected a piece of potash to the poles of a battery and caused the release of a metal at the negative pole. According to his cousin Edmund, Davy began dancing around the room in ecstasy, realizing that he had isolated another element. He named his newly discovered metal potassium after potash.

UNTIL THE LATE eighteenth century, bleaching was accomplished by soaking fabric in buttermilk and then laying it out on the ground to be whitened in sunlight for weeks. These areas, known as bleach fields, took up enormous spreads of land. The nineteenth-century Industrial Revolution created a far greater demand for both soap and bleach. Industry was blackening entire cities, and as skies—and clothes—became covered with soot, it was becoming difficult to find enough space for bleach fields in urban areas.

Another self-taught chemist, a Swede named Carl Wilhelm Scheele, in 1774, twelve years before his celebrated discovery of oxygen, first described a substance called chlorine and noted that it had the ability to bleach. Scheele was also one of the first to study the fermenting attributes of lactic acid.

But it was not until 1786, ten years after Scheele's observations on chlorine, that a practical application was pursued by the great French chemist Claude-Louis Berthollet, who showed that chlorine, when absorbed in potash, created a liquid bleach. Yet another salt-based industry was founded. In little more than a year, industrial bleaching became a major activity in the British textile industry.

In 1810, Davy isolated chlorine and proved that it too was an element, a greenish gas which he named for the Greek word for greenish yellow.

Chlorine has become an important industry. Not only used for bleach, water treatment, and sewage treatment, it is also an ingredient in plastics and artificial rubber. And, as with so many scientific discoveries, a military application was found. Chlorine was the basis for gas warfare. In 1914, at the outbreak of World War I, chlorine gas was exploded in canisters, but later in the war, artillery shells filled with carbonyl chloride proved to be more effective. Known as mustard gas, the compound is credited with 800,000 casualties.

～

CHEMISTS AND ENTREPRENEURS were beginning to understand that "salt" was one of a very specific group of substances that were often found together and that what we now call "common salt" was in many ways the least valuable of the group. In 1744, Guillaume François Rouelle, a member of the French Royal Academy of Sciences, wrote a definition of a salt that has endured. He said that a salt was any substance caused by the reaction of an acid and a base. For a long time, the existence of acids and bases had been known but little understood. Acids were sour tasting and had the ability to dissolve metal. Bases felt soapy. But Rouelle understood that an acid and a base have a natural affinity for each other because nature seeks completion and, as with all good couples, acids and bases make each other more complete. Acids search for an electron that they lack, and bases try to shed an extra one. Together they make a well-balanced compound, a salt. In common salt the base, or electron donor, is sodium, and the acid, or electron recipient, is chloride.

It turned out that salt was a microcosm for one of the oldest concepts of nature and the order of the universe. From the fourth-century-B.C. Chinese belief in the forces of yin and yang, to most of the world's religions, to modern science, to the basic principles of cooking, there has always been a belief that two opposing forces find completion—one recieving a missing part and the other shedding an extra one. A salt is a small but perfect thing.

～

MUCH OF THE new interest in salt, like the early Chinese experiments with saltpeter, was focused on providing the military with ever more efficient ways to blow up people and things. In the nineteenth century, it was discovered that potassium chlorate produced a bigger explosion than traditional gunpowder, potassium nitrate. And magnesium had even more impressive explosive properties.

This science gave birth to a broad range of industries, some of which also poisoned people. The Leblanc process, invented by eighteenth-century French surgeon Nicolas Leblanc, treated salt with sulfuric acid to produce sodium carbonate. Along the way, it also gave off hydrogen chloride fumes and solid calcium sulphide. The calcium sulphide released the classic "rotten egg" smell of sulfur to add to the black clouds and cinder of industrial centers. Hydrogen chloride fumes were worse.

The gas from these manufactories is of such a deleterious nature as to blight anything within its influence, and is alike baneful to health and property. The herbage in the fields in their vicinity is scorched, the gardens yield neither fruit nor vegetables; many flourishing trees have lately become rotten naked sticks. Cattle and poultry droop and pine away. It tarnishes the furniture in our houses, and when we are exposed to it, which is of frequent occurrence, we are afflicted with coughs and pains in the head.—*hearings at the town council of Newcastle upon Tyne, January 9, 1839*

Saltworks, once contaminated by coal smoke and pan scale, expanded their line of products and became far more toxic. By the 1880s, the age of canals had come to an end with the development of railroads, and salt was no longer profitable in upstate New York. But salt was used to manufacture soda ash, caustic soda, bicarbonate of soda, and other chemicals. The salt center of Syracuse was turned into a chemical manufacturing center, temporarily saving the industry but nearly destroying Lake Onondaga with pollution. Chlorine is a component of some of the deadliest industrial pollutants, including polychlorinated biphenyls, which are more infamously known by their abbreviation, PCBs.

On May 15, 1918, the section of the Erie Canal that ran through Syracuse was closed. Five years later, the city bought the canal property for $800,000 and covered it over, creating Erie Boulevard. Soon even the salt industry vanished. In the 1930s, the saltworks were cleared away, and the city struggled to clean up the lake so that the area could be used for recreation.

The canals of downtown Syracuse, New York. Onondaga Historical Association, Syracuse

CHAPTER NINETEEN

The Mythology of Geology

CHEMISTRY CHANGED FOREVER the way we see salt. But it was inventions in other fields that radically changed the role of salt in the world.

Salt for food will never become completely obsolete. But since the beginning of the Industrial Revolution it has steadily become less important.

The first blow was from a Paris cook named Nicolas Appert. Considering the significance of his invention, little is known of Appert. Some think his first name may have been François. He was a confectioner who believed that sealing food tightly in a jar and then heating the jar would destroy the substance that caused food to rot, a substance that he termed *ferment*.

Among the first salt fish customers to be lost to Appert's ideas was Napoleon's navy. In 1803, Appert persuaded the navy to try his broth, beef, and vegetables all preserved in glass jars by his heating and sealing process. The navy was pleased. A report stated, "The beans and green peas, both with and without meat, have all the freshness and flavor of hand-picked vegetables."

Anyone who has ever eaten canned beans or peas may suspect some hyperbole here, but for sailors who had never had vegetables on their long voyages, Appert's treats seemed a wondrous invention. Grimod de La Reynière, the leading gastronomic writer of France at the time, praised Appert's food.

Appert's 1809 book, *The Art of Preserving All Kinds of Animal and Vegetable Substances for Several Years*, was widely read and even translated into English. Only months after its publication, Peter Durand, a Londoner, was granted a patent for preserving food. He admitted that his ideas came from an unnamed foreigner, who was probably Appert. Actually, in 1807, an Englishman named Thomas Saddington had demonstrated a similar process. But what is important about Durand is that along with glass and pottery, he mentioned in his list of possible containers for preserved foods "tin and other metals."

Bryan Donkin, a visionary early British industrialist, realized, perhaps better than Durand had, the potential of the tin idea. He had founded the Dartford Iron Works, and, in 1805, he helped finance the first industrial papermaking machine. After Durand received his patent in 1809, Donkin founded Donkin, Hall, and Gamble, the first British canning plant, across the Thames from the City of London. It became the outfitter of famous expeditions such as the arctic expeditions of William Edward Parry in the 1820s.

Toward the end of the Napoleonic Wars, the British navy began experimenting with canned food from Donkin, Hall, and Gamble. At first, canned food was used as special provisions for those on sick list, but by the 1830s, it had become part of general provisions.

Unfortunately, the can opener had not yet been invented. Sailors were issued special knives with which to pry open the cans.

In 1830, a canning plant was built in La Turballe, the sardine fishing town across the opening of the Guérande swamp from Le Croisic. The plant flourished, and gradually most of the area's salt fish business collapsed, unable to compete with canned products. In time, much of the French Atlantic salt fish industry disappeared. A similar fate befell much of the salted herring industry to the north and anchovy industry to the south.

A TWENTIETH-CENTURY invention dealt an even worse blow to the salt fish industry and, for that matter, to fish. The idea of using cold to preserve food had been much thought about in the nineteenth century. In 1800, Thomas Moore, an American engineer who wanted to keep his butter cool during the twenty-mile journey between his Maryland farm and the market in Washington, D.C., the newly created capital, built a wooden box with a metal butter container inside surrounded by ice. He then stuffed the box with rabbit fur. According to his account, his butter, firm and chilled even in summer, sold well in Washington.

As early as the 1820s, fish was sometimes packed in ice in an attempt to preserve its freshness. American farmers asked themselves if ice could not somehow be used like salt. "Salting in snow" was discussed by Sarah Josepha Hale. As the editor of *Godey's Lady's Book* from 1837 to 1877—years in which this widely read magazine almost never mentioned the Civil War because war was not the business of ladies—Hale was regarded as one of America's most influential women.

An excellent way to keep fresh meat during the winter, is practiced by the farmers in the country, which they term "salting in
snow." Take a large clean tub, cover the bottom three or four
inches thick with clean snow; then lay pieces of fresh meat,
spare ribs, fowls, or whatever you wish to keep, and cover each
layer with two or three inches of snow, taking particular care to
fill snow into every cranny and crevice between the pieces, and
around the edges of the tub. Fowl must also be filled inside
with the snow. When the tub is filled, the last layer must be
snow, pressed down tight; then cover the tub, which must be
kept in a cold place, the colder the better. The meat will not
freeze, and unless there happen to be a long spell of warm
weather, the snow will not thaw, but the meat remain as fresh
and juicy when it is taken out to be cooked, as when it was first
killed.—*Sarah Josepha Hale,* The Good Housekeeper, *1841*

An eccentric New Yorker named Clarence Birdseye was troubled by the idea of packing food in ice, because the ice melted
and the resulting water created an environment in which bacteria
could flourish. Bored with New York office jobs, Birdseye had
moved to Labrador with his wife, Eleanor, and their son to earn
his living trapping furs. He noted, as was long known by the indigenous people there, that when fish are caught in Labrador in
the wintertime, they instantly freeze, and that if kept this way for
several weeks, when thawed they will taste fresh.

The Birdseye home became very different from other Labrador
households. Cabbages were frozen in the windows, and fish were
swimming in the bathtub, as Birdseye experimented. He observed how the harsh Labrador wind acted on wet food, freezing
it very rapidly so that bacteria had no opportunity to develop.
Soon he was in Washington, unveiling his new technology, the

fast-freezing process. Birdseye went to the unveiling equipped with a block of ice, a fan, and a bucket of brine—all the necessary ingredients for a homemade Labrador winter. He made the brine from calcium chloride, which, after experimentation, he found kept the temperature lower than sodium chloride.

Fast freezing worked, Birdseye discovered, because of a principle every salt maker knew: Rapid crystallization creates small crystals, and slow crystallization produces large ones. Because the ice crystals in rapidly frozen food were small, they did not interfere with the tissue structure and so better preserved the food in its original state.

In 1925, Birdseye moved to Gloucester, Massachusetts, the leading New England cod-fishing port, and established a frozen seafood company. Birdseye's invention came at a time when the demand for salted fish was in rapid decline in both the United States and Britain. The railroad, faster transportation, and better market systems had introduced more people to fresh fish. By 1910, only 1 percent of the fish landed in New England was cured with salt.

By 1928, 1 million pounds of food frozen in the Birdseye method was being sold in the United States. Most of it was being sold by Birdseye, who managed to find a buyer for his company just before the 1929 market crash. The company became General Foods, modeling the name after General Electric and General Motors, leaders in their respective industries. Birdseye once said, "I do not consider myself a remarkable person. I am just a guy with a very large bump of curiosity and a gambling instinct." By the time he died at age sixty-nine, he had patented 250 inventions including dozens of devices and gadgets to improve the operation of his frozen-food process. He invented a lightbulb with a built-in reflector and a gooseneck lamp. But he will always be remembered for frozen food.

Fast freezing had at last made the unsalted fish people wanted, available to everyone, even far inland. Soon fishing vessels, instead of salting their catch at sea, were freezing it on board. Most salted foods became delicacies instead of necessities.

⌒

THE AGE OF industrial engineering brought inventions to a salt industry that had been slow to develop new ideas. Most saltworks had been started as small operations by individualists who found original solutions to their technical problems. Some ideas, like the natural gas of Sichuan, had enduring and far-reaching applications. Some ideas, such as little paddle wheels with bells in the freshwater canals of Lorraine saltworks, to ensure by their tintinnabulation that the canals were not mixing with brine canals, were purely local. Still other ideas were based on cheap—often family—labor. One of the more curious examples of this was the *grau*, a sixteenth-century machine for lifting brine from storage tanks by means of a basket on one end of a lever. On the other end were ropes. Women would grab the ropes and swing off them like children at an amusement park, their weight hoisting the buckets on the other end.

Another example, using cheap labor, in this case slave labor, was a human-powered wheel used to pump brine. In medieval Salsomaggiore, men, chained at the neck, walked on the slats of huge wheels as on a treadmill. In Halle, brine was lifted on a wheel powered by twelve men. The man-wheel was used in Europe until the nineteenth century. In 1840, a twenty-eight-pond saltworks near Cape Ann, Massachusetts, supplemented the power from windmills on calm days by pumping brine by means of a fifteen-foot-in-diameter, five-foot-wide wheel with buckets on its outer rim. The wheel was powered by a large bull that walked inside the wheel.

Pumping brine was one of the most important engineering problems confronting salt makers, and it inspired many inventions. The first engine, the steam engine, which led the way to the Industrial Revolution, was invented in 1712 by an Englishman, Thomas Newcomen, and used exclusively for pumping water. The engine and its subsequent improvements were embraced by British and American salt makers, who had abundant fuel, mostly coal. In Germany, however, where there was not enough sunlight for solar evaporation and most of the springs had relatively weak brine, the cost of fuel was the central problem. In the seventeenth century, the Germans, learning that the salt at Salsomaggiore was more profitable than that in Germany, had sent investigators to Parma, convinced they would find new fuel-economizing technology. Instead they found that the Parmigianos simply charged a great deal more for their salt. For the Germans, steam engines consumed too much fuel.

⌒

SALT INSPIRED INNOVATIONS in transportation, perhaps none more impressive than the canals of northern Germany, Cheshire, and the United States. The Anderton boat lift lowered entire loaded salt barges fifty feet from the Cheshire canal system down to the level of the River Weaver, which ran into the mouth of the Mersey across the bay from Liverpool. Built in 1875 to link the Trent and Mersey Canal to the River Weaver, it originally lowered the barges by a cantilevered hydraulic system based on counterweights and water power. But salt spills eventually turned the canal brackish and corroded the machine. In the twentieth century, an electric motor was added.

But it was in the technology of drilling, that salt producers had a momentous impact on the modern world. For a long time, the percussion drilling techniques of the Chinese were the leading invention. All percussion drilling, from early Sichuan to nineteenth-century Kanawha, essentially consisted of a chisel with a long shaft being whacked by a kind of hammer. In the sixteenth and seventeenth centuries, Europeans began using a rotary drill. They attached extension rods, known as boring rods, which, in 1640, enabled the Dutch to drill 216 feet under Amsterdam to reach a source of fresh water.

In the early nineteenth century, the drilling proved so successful at Kanawha that many Americans began deep-drilling projects in search of salt. An improved connection between the driving shaft and the drill shaft was developed in the United States; this connection was called a jar because it was designed to better withstand the jar of the pounding shaft. Europeans quickly adopted the American invention. Jars had actually been used centuries earlier by the Chinese, but westerners did not know this. At the time of the American invention of the jar, a western missionary, one Father Imbert, had gone to China to study the ancient wells of Sichuan. He reported on more than 1,000 ancient wells drilled to great depths and brine lifted in long bamboo buckets. He also observed that the Chinese had elaborate techniques for recovering broken drill shafts. In the West, such obstructions were often the cause of a well being abandoned.

⌒

IN THE LATE seventeenth century, when coal prospectors drilled into the Cheshire earth and found rock salt, it was the scientists, not the salt merchants, who were excited by the find. It demonstrated how improved drilling might someday open up an entirely new scientific field—geology, the study of the earth. Almost another century and a half would pass before England had its first systematic curriculum in the study of geology—established not by a geologist but by Humphry Davy.

Long before there were geologists, there were natural philosophers who contemplated the structure of the earth. Some of their best ideas remained unproved and unembraced. Nineteen hundred years before Columbus's voyages, Aristotle wrote that the earth was round. An eleventh-century-A.D. Persian physician, Avicenna, author of some 100 works on medicine and philosophy, wrote about land being formed by prehistoric flooding, erosion, sediment deposits, and the metamorphosis of soft rock. He might have been remembered as the father of geology if more people had understood what he was talking about. But it would take centuries for the scientific world to catch up with him.

Throughout the Renaissance, new ideas were presented on the earth's formation by thinkers in various fields, including Leonardo da Vinci, who opined that fossils were not, as widely supposed, placed in the rock by the devil but were formed by trapped plants and animals metamorphosing in the soil.

In the mid–sixteenth century, Georg Bauer, a German with the pen name Georgius Agricola, wrote on the origin of mountains, minerals, and underground water. His 1556 work *De re metallica* was the most complete work to date and for centuries to come on techniques for mining and producing metals and minerals, including salt.

Long before it was called geology, a number of geological debates persisted. One of them was on the origin of salt. Was a giant bed of salt at the bottom of the sea keeping ocean water salted? Or, as some believed, did the tremendous pressure at great depths so squeeze water that it turned salty? Another theory held that salt did not come from the ocean at all, but that salt on earth was carried to sea by rivers.

In the seventeenth century, René Descartes asserted that sweet water was soft and would evaporate, but salt particles were hard and would remain, and that was why the sea remained salty. According to his theory, the soft part of the ocean, the freshwater, was absorbed in the earth's pores and then reappeared in the form of freshwater rivers, streams, and lakes. The earth not only had pores, but also had cracks, and these fissures were wide enough to let in the seawater, particles and all. This seawater usually formed brine springs. But some of these fissures were dead ends and did not lead to springs. The seawater that seeped into such places hardened into rock salt.

One eighteenth century theory held that the source of natural brine was that gypsum saturated with seawater leached salt. But another theory was that gypsum, a soft mineral common in most of the world, turned into salt. Water, according to this hypothesis, is salty in its natural state. The real question was: What caused freshwater not to be salty?

Robert Hooke, the seventeenth-century philosopher whose many scientific accomplishments include originating the word *cell* for the basic organism, concluded that salt came from the air. Others concluded that salt came from alkali, which turns out to be true, since alkali are bases. Some combined the two, concluding that salt was caused by the alkali in seawater mixing with the salt in the air.

The Germans tried to understand their many brine springs. Did brine come from rock salt below, as already appeared to be the case in Cheshire? Christian Keferstein, a Prussian lawyer, self-taught scientist, and author of a seven-volume geologic study, was convinced that the discovery of rock salt near a number of brine springs was coincidental. Rock salt, he believed, came from certain rocks.

In the eighteenth and nineteenth centuries, the raging geologic debate pitted neptunism against plutonism. The neptunists, led by German mineralogist Abraham Gottlob Werner, believed that the source of all bedrock was a common ancient sea. According to plutonism, most rock had hardened from a huge molten rock mass. Neptunism held that salt came from the sea, and plutonism insisted it was volcanic in origin.

In 1775, William Bowles used the salt mountain of Cardona to argue against the neptunism theory of salt. Logic indicated that such a huge mountain of solid rock was probably not left over from the ocean. Several others confirmed that this Pyrenees-sized mountain was solid salt or mostly salt—70 percent, one study contended—and that such a mass must have metamorphosized out of other rock. Eventually, neptunism was rejected, because both granite and basalt were proved to be of volcanic origin. But did that mean that plutonism was right about salt being formed by volcanos?

In the nineteenth century, Europeans became extremely curious about the structure of salt deposits in other parts of the world, such as the Dead Sea. Thomas Jefferson was constantly questioned by Europeans about American salt formations, some of them mythical structures rumored to exist in the wilderness of the northern reaches of the Louisiana Purchase.

As drilling improved, it became clear that the earth possessed huge underground salt deposits, and that far from being rare, rock salt was very common. By the end of the eighteenth century, many geologists were convinced that most of central Europe was sitting on an enormous salt deposit. To a large extent, they were right. The same salt deposit that feeds Alsace and Lorraine extends through Germany to the Austrian Salzkammergut. The thick layer of salt underneath Cheshire starts in Northern Ireland and runs into northern Europe. Onondaga County, New York, is part of an enormous salt field that stretches across the entire Great Lakes region, providing rock salt mines under the city of Detroit, in Cleveland, and in Ontario.

There is still not complete agreement on the formation of many of the earth's great salt deposits. But they are generally agreed to have had their origin in oceans rather than volcanos, though there is still no set explanation for the saltiness of the sea.

Geologists, both out of curiosity and in search of salt, looked for salt domes, areas like Avery Island that had deep pure sodium chloride deposits forced by the pressure of shifting plates to mushroom up from the depths and break the earth's surface in a dome shape. The history of salt dome theories begins with Thomassy, the Frenchman who said that Avery Island was rock salt. He stated that the salt there "comes from a volcano of water, mud, and gas." This plutonistic theory of the salt's volcanic origin was later rejected by most geologists.

In 1867, C. A. Goesmann, reporting to the American Bureau of Mines, theorized that the salt under Avery Island resulted from brine springs ascending through older deposits of bedded salt. According to Goesmann, brine rises from deep within the earth, moving through the earth's fissures and crystallizing near the surface.

Salt prospectors were able to find salt domes by recognizing the rounded shape that protruded above the earth's surface. They would drill in such spots, which invariably yielded brine, but often the salt was so contaminated with blackish muck that it was of little commercial value.

In 1901, two men, Pattillo Higgins and Anthony Lucas, ignored the advice of geologists and started drilling a Texas salt dome called Spindletop. No one ever looked at salt domes the same way again. No longer were terms like *well* and *drill rig* to conjure up the image of salt. Spindletop had spawned the age of petroleum.

Such an age had been promised in 1859, outside Titusville, Pennsylvania, where Edwin Drake, after studying the drilling techniques of salt producers, drilled 69.5 feet and, to everyone's surprise but his, hit oil. He began producing twenty-five barrels per day, and many started to believe that oil would be an abundant U.S. resource. But subsequent drilling, mostly in the East, yielded little.

By 1866, seven years after Titusville, when salt was discovered in Ontario, it was a different age. Canada had not produced much salt, but instead of excitement about a rich new salt field, there were high hopes that oil had been found. In Goderich, Ontario, Samuel Platt organized the Goderich Petroleum Company, which began work on the north bank of the Maitland River—drilling 686 feet through gray limestone. There was no sign of oil, and the stockholders who had provided $10,000 in start-up money wanted to abandon the project. But the county council offered Platt a bonus of $1,000, and the city offered $500 provided he continue to a depth of 1,000 feet. At 964 feet from the collar of the hole, he hit solid rock salt.

The Goderich Salt Company was founded with fifty-two boiling kettles, and the Ontario salt fields have become one of the most productive saltworks in the modern world.

⌒

BY THE TIME Higgins and Lucas began drilling at Spindletop, hopes for American oil had faded. But Spindletop changed the thinking of geologists, chemists, engineers, and economists because it showed that a single spot, a corner of a single salt dome, could by itself produce enormous quantities of oil in a short period of time. In its first sixty-five years, Spindletop produced 145 million barrels of oil. As a result of Spindletop, the United States surpassed Russia, the largest oil producer at the time.

Also because of Spindletop, geologists took a new look at salt domes. Because salt is impenetrable, organic material gets trapped next to the salt and slowly decomposes into oil and gas. For this reason, oil, gas, or both are frequently found on the edge of salt. The 2,000-year-old mystery of Sichuan was answered.

After Spindletop, more oil was found along the Texas-Louisiana coast in such places as Sour Lake in 1902, Humble in 1905, and Goose Creek in 1908. The United States took the lead in a drilling technology that was now in demand all over the world, as geologists searched the globe for likely salt domes to drill. Many of them were found in the Persian Gulf. In 1908, oil was found in Persia, now Iran, in the places where Herodotus had written about salt.

Exploration continued in North America. Few believed Columbus Joiner when he began drilling for oil on an unheard-of geologic structure he called "the Overton anticline." It is now known that his theory of geology was completely wrong. But fortunately, at the time, no one could disprove it. They laughed at him, and he drilled anyway and found the largest oil field in North America, the East Texas Field.

In a less corporate age, oil men used to take glee in pointing out that the three most important discoveries in the history of American oil—Titusville, Spindletop, and the East Texas Field—were all drilled against the advice of geologists.

As Brownrigg had predicted in the mid–eighteenth century, "Old arts are improved and new ones daily invented." The quest for salt had turned unexpected corners and created dozens of industries.

CHAPTER TWENTY

The Soil Never Sets On . . .

WHEN THE BRITISH Empire was at its height, "Liverpool salt" was the salt of the empire, a prestigious product known all over the world. As in Cardona, Hallein, and Wieliczka, a visit to the Cheshire salt mines was a special treat for visiting aristocrats. These elite guests were lowered into the mines in enormous brine buckets. The candlelit bucket passed through the narrow shaft and when it came out at the mine below, the visitors were greeted by the word *welcome* spelled out by the workers with candles on the salt floor. According to local legend, when the czar of Russia visited England, he dined beneath Cheshire by the "light of a thousand candles."

It was the canals leading to Liverpool that had given Cheshire a global market. Not only was salt ballast for the voyage to America to pick up cotton and other imports for British industry, but, because the port of Liverpool was deeply involved in the slave trade, ships regularly bound for West Africa needed an outbound cargo. Nigeria bought Cheshire salt until 1968, when that market collapsed with the Biafran civil war.

By 1890, besides the lucrative foreign market, Cheshire supplied 90 percent of British salt. In Cheshire, a good income could be had by anyone who could buy or lease a small plot of land near one of the wiches and who had the relatively small amount of capital needed to drill a hole in that ground and set up some wide, flat iron pans over a coal-burning furnace.

Because the chimneys at Cheshire brine works were not built high enough for the wind to carry the soot and glowing cinders away, workers and townspeople lived amid burning black clouds—"the smoke and smother of weary Winsford," as one newspaper described it in the 1880s. An 1878 royal commission

Tourists being lowered into a nineteenth-century Cheshire salt mine.
The Salt Museum, Cheshire County Council, Northwich

reported that air pollution was choking the local vegetation. Salt producers were fined for the pollution, but this did not alter their practices. One producer told a board of inquiry that he would continue until the fines drove him out of business, and then he would relocate elsewhere.

Salt did not provide an easy life. Often a man would rent a pan, and he, his wife, and children would ensure the maximum profit by working around-the-clock shifts to keep the pan in constant production. Children would start working at the salt pans at the age of nine. Women would go back and forth between the pans and their homes, alternating household work with salt making. A normal workday for a salaried salt worker was twelve hours, but it was often much longer. Some were paid by the hour and others by the quantity of salt they produced.

Reform came slowly. An 1867 law forbade women and children to work between 6 P.M. and 6 A.M. Factory inspectors began protesting the working conditions for women, saying the work was too physically strenuous. And the public was scandalized when inspectors revealed that in the hot boiling houses, men and women worked together, the men stripped to the waist and the women, dresses removed, in their underwear and petticoats.

An 1876 inquiry demanded that girls under eighteen be barred from saltworks. Robert Baker, one of the inspectors, argued for shorter workweeks for men. He told the board of inquiry, "The fact is the men never see beds but on Saturday night."

For centuries, Nantwich had been the leading Cheshire salt town. But in the early twentieth century, geologists discovered that the most important deposits were under Northwich and

Winsford, where the thickness of the rock salt is as much as 180 feet and even at its thinnest, no less than 48 feet.

The strange sinkholes that had been sporadically appearing in the eighteenth century had become by the late nineteenth century a regular phenomenon—not so much in Nantwich, but in and around Northwich and Winsford. Every year new spots in meadows, pastures, and even towns collapsed. The holes caught rainfall and made small lakes. Toward the end of the century, a lake of more than 100 acres suddenly appeared near Northwich. Sometimes saltworks made use of the newly developed holes, filling them with ash or lime waste, just one more pollutant in an area black with coal smoke.

The brine makers tried to continue blaming the sinkholes on the rock salt miners, saying sinking was caused by abandoned mine shafts. This had worked better when rock salt was a new discovery. But in the nineteenth century, it became obvious that

Subsidence: office buildings sinking in Castle, Cheshire. The Salt Museum, Cheshire County Council, Northwich

the location of the sinkholes bore no relation to the location of mine shafts, and as sinking became more frequent, there were not enough shafts to explain the number of occurrences. On the other hand, there was an exact correlation between the increase in brine production and the increase in sinkholes.

The sinking was starting to wreak havoc with railroad lines and even to threaten bridges. In Northwich and Winsford, homes and buildings collapsed as the ground gave way underneath them. By 1880, 400 buildings had been destroyed or damaged in Northwich alone. At Winsford, a new church was condemned as dangerous. Water mains, sewer lines, and gas pipes were continually breaking, and the costs of repairing them were draining municipal budgets. Shop after shop was condemned and torn down.

A passing traveler described Northwich:

A number of miniature valleys seem to cross the road and in their immediate neighborhood the houses are, many of them, far out of the perpendicular. Some overhang the street as much as two feet, whilst others lean on their neighbors and push them over. Chimney-stacks lean and become dangerous; whilst doors and windows refuse to open and close properly. Many panes of glass are broken in the windows; the walls exhibit cracks from the smallest size up to a width of three or four inches; and in the case of brick arches over doors and passages, the key brick has either fallen out or is about to do so, and in many cases short beams have been substituted for the usual arch. In the inside, things are not much better. The ceilings are cracked and the cornices fall down; whilst the plaster on the walls and the paper covering it, exhibit manifold chinks and crevices. The doors either refuse to open without being continually altered by the joiner, or they swing back into the room the moment they are unlatched.—Chamber's Journal, *1879*

With an English flair for genteel euphemism, the growing disaster was labeled *subsidence*. Subsidence in Cheshire was becoming a subject of considerable amusement around England, spawning Cheshire jokes. But it also drew religious fanatics, who went to Cheshire to deliver sermons to the crowds who came to gawk at the holes. The preachers would stand at the edge of the craters looking down into akimbo boiling houses and broken smokestacks and warn that this was what hell would look like.

The truth was that too much brine was being pumped too rapidly from underneath Cheshire. Hundreds of ambitious small-scale entrepreneurs were making salt. They became extremely competitive. Some would pump additional brine out and dump it in the canals just to try to deprive their competitors.

The brine that flows over the salt rock of Cheshire is a saturated solution—one quarter salt—and so it is incapable of absorbing any more. But as brine was removed, fresh groundwater took its place, and this water would absorb salt until the brine was once again one-fourth salt. The problem was that if large quantities of brine were removed, they were replaced with large quantities of freshwater that hungrily absorbed considerable amounts of salt. Once that started happening, the freshwater began eroding the natural salt pillars that supported the space between the salt rock and the surface. When a pillar collapsed, the earth above it sank.

But even in the nineteenth century, when this process was understood, it was difficult to know whom to blame. The area around a saltworks might remain solid even though the brine it was pumping was causing the earth to collapse four miles away.

Two or three other saltworks, though closer to the hole than the culprit, might have caused no damage at all.

Identifying the culprit was an important legal issue, since hundreds of people, many of them not in the salt industry, had lost their property and were demanding compensation. Unable to name a defendant, they could not pursue a legal action. Could they charge the salt industry in general? Citizens formed committees and went to Parliament proposing a bill that compensated victims for the damages caused by the salt industry. Property owners, citing a long-standing principle of British law that the owner of land owned the subsoil, claimed that not only was their property being destroyed, but they were being robbed of the rock salt that they owned. The brine pumpers were sucking up their rock salt from under their own sinking property.

The salt producers argued, with typical nineteenth-century capitalist confidence, that the locals were already being compensated by the economic benefits of having the salt industry. They denied that the subsidence was caused by pumping, insisting that the sinkholes were a natural phenomenon that would continue even without pumping. These arguments prevailed, and, in 1880, the bill was defeated.

⌒

IN 1887, A group of London financiers raised £4 million to buy up saltworks for a company called the Salt Union Limited. The company, founded by seven entrepreneurs without prior connections to the salt industry, wanted to buy up all British salt production and become the largest industrial company in England. Both the London *Times* and the *Economist* warned that such a giant could not maintain a monopoly on an industry whose raw material was so common and initial investment requirements so modest.

In Cheshire, with its long tradition of individualists and small private operators, many were angered at the sight of a corporate giant buying out local salt makers one by one. But industry leaders felt that the Salt Union was a workable solution to a sector that had too many participants. The rate of brine pumping spurred by this competition was in danger of literally sinking them all. The low salt prices of the late 1880s gave a further incentive to selling out.

Sixty-five salt producers sold out to the Salt Union. They were not only from Cheshire but from neighboring Staffordshire, Worcestershire, northeastern England, and Northern Ireland. The Salt Union had cornered 85 percent of British salt production. But most analysts believed that it had greatly overpaid to acquire these companies.

Nevertheless, the company was highly profitable its first few years, before going into a steep decline. Not until 1920 did profits again reach the level of 1890.

In 1891, when the Cheshire Salt Districts Compensation Bill again came before Parliament, the Salt Union used the arguments that the independent salt producers had used a decade earlier: that the people were being compensated by the economic benefits of having a salt industry and that the sinking was a natural phenomenon that would have occurred without saltworks. But now the Salt Union provided a target, a single entity that was clearly responsible—a defendant. The local citizenry spent a fortune promoting the bill, and both the Salt Union and its shareholders spent a fortune fighting it. It passed, though, and within ten years the Salt Union itself was applying for damages, saying its properties had suffered subsidence from the pumping of others.

In the long run, the Compensation Act probably helped the Salt

Union. It created the Cheshire Brine Subsidence Compensation Board, which was financed by a flat tax on salt producers. The cost of the tax was onerous for small operations and insignificant for large ones. The small-scale producers regarded the Brine Board, as it was known, as another attempt by big salt to drive them out.

The Brine Board established a building code that had to be followed for new buildings to be eligible for compensation. The collapsing towns were rebuilt in the old Tudor style, with each new house resting on a timber frame that had built-in anchors for the placement of hydraulic jacks powerful enough to lift sinking buildings. An eighteen-inch lift looking like a canister was capable of hoisting fifty tons. It was the same technology that had been used to lift salt barges at the Anderton boat lift.

ENGLAND IS SOMETIMES thought of as a land of eccentrics who stubbornly cling to quaint and hopelessly outmoded ways, but it is also the land of entrepreneurs who created the Industrial Age. British industrialists built powerful companies, such as the Salt Union, that were the forerunners of today's multinational giants. In Cheshire, these two kinds of Englishmen were represented by the Thompsons and the Stubbses.

Both families have long histories in Cheshire salt. A 1710 map marks "John Stubbs salt pit," though today's Stubbses do not know exactly who John was. Some Stubbses were dreamers. It was a Cheshire Stubbs who built a plantation in the Turks and Caicos Islands. But in the nineteenth century, the Stubbses joined the Industrial Age and sent their sons to school to study engineering.

The Thompsons were cursed with longevity. So while the Stubbses' family operations were run by well-educated young engineers schooled in new technologies, the Thompsons' family salt business was often run by octogenarian grandfathers and great-grandfathers.

Both families had saltworks near the sinking town of North-wich. Eventually, various members of the Stubbs family had salt-works all over Cheshire. But toward the end of the century, brine works, unable to compete with large companies, were one by one going broke, sinking either financially or literally. In the 1870s, various Stubbs brothers consolidated their operations, and in 1888, they sold out to the Salt Union.

After selling out and taking a seat on the board of directors, some of the brothers opened new saltworks across the county line. Then, in 1923, they bought the New Cheshire Salt Works near Northwich.

The Thompsons were not that different. In 1856, they started the Alliance Salt Works by digging a hole behind the Red Lion Hotel. But they too wanted to remain an independent family op-eration, and after they sold the Alliance Salt Works to the Salt Union in 1888, they dug a new shaft beside the Red Lion Hotel and called it the Lion Salt Works. New technology had concen-trated on finding salt and bringing it to the surface. But once the brine reached the saltworks, little had changed since the time of the Romans. It was still evaporated in lead pans. The pans had gotten larger than the three-by-three-foot Roman ones. The Thompsons had thirty-by-twenty-foot lead pans, heated by coal that was stoked from four furnace doors in the huge coal oven under the pan. Into the nineteenth century, even the pipes for brine were still made from hollowed tree trunks. Except for larger pans and coal being burned instead of wood, most of the process was described in Georgius Agricola's 1556 work *De re metallica*, which remained the standard European text on salt making. The work was first translated into English in 1912 by mining engineer and future U.S. president Herbert Hoover and his wife, Lou Henry Hoover.

Cheshire salt makers had become skilled at making different products for an expanding and varied list of customers. Most of this simply involved adjusting the cooking time. Dairy salt was cooked fast, so that the rapidly moving water would create fine crystals used in butter and cheese. The large grains of solar-heated sea salt were replicated with the slow heating of so-called fourteen-day salt, which was shipped to Grimsby for salting cod. Salt hardened into blocks and then crushed was locally called "Lagos salt" because it was shipped to West Africa. Because the West African market bought salt by volume rather than by weight, all the Cheshire companies made a large, lightweight crystal for that market.

But in 1905, James Stubbs went to Michigan to learn about a new "evaporator." The fundamental concept of a vacuum evaporator is that lower pressure reduces the boiling point of a liquid. A boiler produces steam, which heats a chamber, an evaporator. The steam is then piped into a second evaporator. The second evaporator cannot heat to as high a temperature, but because it is in a vacuum, the pressure is lower and less heat is needed for steam. This steam can then be passed to a third evaporator. And so an entire series of evaporators can be operated on the fuel that was expended to heat the first one. This solves one of the oldest problems in salt brine production, the problem the ancient Chinese solved with gas—the cost of fuel.

Liverpool sugar refiners had been using steam evaporators since 1823, when William Furnical had introduced the use of steam heat in sugar refining. In 1887, the first vacuum pan salt process was put in operation by Joseph Duncan in Silver Springs, New York. The evaporator heated brine to steam and forced it

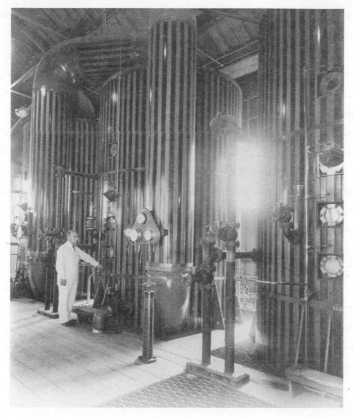

The Stubbses's first vacuum evaporator. New Cheshire Salt Works, Ltd.,
Northwich

into a tank, where salt crystals formed; once the crystals reached
a certain size and weight, they dropped through the bottom. If
they were too big, they would be washed back up by the incom-
ing brine; if too small, they would not drop down. For the first
time in the long history of salt, a salt was being made in which
every crystal was the same size.

The steam from the first tank was used to heat a second tank and a third tank. Today, up to six or even eight tanks, each evaporating brine, can be fueled by the first evaporating tank.

In the 1930s, the Stubbses finally imported their first salt evaporator to the New Cheshire Salt Works. This magnificent triple-towered machine was art deco in design, with vertical stripes of dark and blond wood and polished brass fittings and gauges. For a while, the Stubbses still had some open pans for larger crystal salt, but in time, the old pans could no longer compete economically with modern evaporators. New, more efficient evaporators were bought in the 1950s and again in the 1990s. The Stubbses, along with the Salt Union, are among only three surviving commercial British salt producers.

Anyone who makes bread in any quantity finds himself getting through a deal of salt. What salt then to use? Since with the exception of the famous Maldon salt from Essex on the east coast—again a luxury—there is now no sea salt extracted from English or Scottish waters, I use Cheshire rock salt sold in 1½ pound blocks or 2 pound bags or 6 pound clear plastic jars, the latter being the best value and the most convenient. This salt is produced by the old Liverpool firm of Ingram Thompson (mention of "Liverpool salt" occurs quite often in eighteenth- and nineteenth-century cookery books) whose salt works are at Northwich. This firm's packaging is minimal and their wholesale prices fair, so if you find you are paying too much for rock salt or "crystal" salt it is probably because middlemen have bought it in bulk and are charging retailers more than is fair.
—*Elizabeth David,* English Bread and Yeast Cookery, *1977*

WITH BRINE WORKS closing or sinking all around them, the Thompsons continued in the old ways. They sent salt blocks to Cheshire schools for children to make salt carvings. At Christmastime, workers still dipped branches into brine pans to grow crystals. In the 1960s, they were still employing hand riveters to mend their lead pans and a steam engine to bring up the brine. Then, in the late 1960s, Nigeria, the Thompsons' principal remaining market, was destroyed by the Biafran war.

Many artisans have been faced with the choice of whether to industrialize or remain a small shop. But at a certain point that choice can be lost. If the operation becomes too unprofitable, it will no longer be able to attract the investment needed to modernize. This was the fate of the Thompsons. They hung on for more than a decade without making money; finally, in 1986, they gave up.

CHESHIRE IS NOW green English countryside. The pastureland is spotted yellow and purple with wildflowers and hedges over which blackberry vines twist. Reedy swan's paddle grows in the unused canals. It is hard to believe that 100 years ago the sky was black with coal smoke, the horizon filled with a hundred smokestacks, the soil contaminated, arid, barren, and scarred white where the pan scale was dumped.

Local people write to the Thompson works and ask for the old salt, which was packed into wooden tubs and dried into salt blocks. They call it lump salt and say it is better for cooking beans and for curing meat. But there is no more Thompson salt.

The Vale Royal Borough Council bought the site and established a charitable trust, which is trying to make this last of the

Northwich brine works a restored museum. It is struggling to get funding because sinkholes surround it. Black-and-white cows lazily nestle into the sinkholes, which are covered with grass and shrubs and delicate white Queen Anne's lace. But every now and then, a little more subsidence is seen, something else sinks. Many believe that one day the old Thompson brine works will sink too.

CHAPTER TWENTY-ONE

Salt and the Great Soul

TWENTIETH-CENTURY INDIA lived under the kind of colonial administration Madison and Jefferson had rejected—the kind that would have made Adams angry. And it did anger a great many Indians. To the British, the Indian economy existed for the enrichment of Great Britain. Industry was for the profit of the English Midlands. Indian salt was to be managed for the benefit of Cheshire.

That the British always saw India as a commercial venture is demonstrated by the fact that when they first gained a foothold on the Asian subcontinent, they placed it in the hands of a private trading company, the Honourable East India Company. Founded in 1600 by a royal charter granted by Elizabeth I, the East India company, though a commercial enterprise, could function as a nation, minting its own money, governing its employees as it saw fit, even raising its own army and navy and declaring war or negotiating peace at will with other nations, providing they were not

Christian. The company bought its first Indian property in 1639, a strip of coastline, and by the end of the century was building the city of Calcutta. A series of eighteenth-century battles between the British and the French eventually secured India for the British, who turned most of it over to the British East India Company.

The company established a sophisticated bureaucracy with a large, well-paid civil service. No high-ranking post was ever given to an Indian. By the nineteenth century, more than half of India was governed by the East India Company and the rest by local princes, who served as puppet rulers for the British. In 1857, Indians openly revolted, and the following year, once the British army had put down the rebellion, the British Crown took over most of the local government from the East India Company.

⌒

BEFORE THE BRITISH created artificial trade barriers, India had affordable, readily available salt. While it has huge saltless regions, with natural salt fields on both its coasts and huge rock salt deposits and salt lakes in between, India had an ancient tradition of salt making and trading. Although the extensive rock salt deposits in Punjab are unusually pure, strictly religious Hindus have always had a distrust of rock salt and even salt made from boiling. Indians have always preferred solar-evaporated sea salt not only for religious reasons but because it was more accessible. On the west coast, by what is now the Pakistani border, and on the east coast near Calcutta, river estuaries spread out into wetlands and marshes where the sun evaporates seawater, leaving crusts of salt.

On the west coast, in Gujarat, salt has been made for at least 5,000 years in a 9,000-square-mile marshland known as the Rann

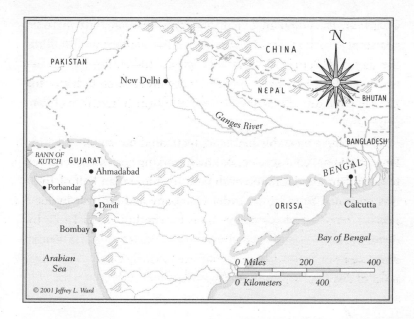

© 2001 Jeffrey L. Ward

of Kutch. This marshland is covered by the sea and flooded rivers in the rainy season from August to September; in December, the salty water begins to evaporate with help from dry wintry winds from the north.

On the east coast is a salt-producing area known as Orissa, with a perfect natural sea salt zone along a tract that is 320 miles long and ten to sixty miles deep. The salt beds, called *khalaris*, are flooded in two spring tides, which saturate the soil with salt as the water evaporates. Salt made from natural solar evaporation was called *kartach*. A second salt, *panga*, was produced by mixing salty soil in seawater and boiling it. The salt was a permanently renewed resource, which rendered this stretch of land not only ideal for salt making but useless for anything else. In Orissa, the poorest of peasants could make salt on the khalari to use or to sell.

The salt makers would clear the field, the khalari, of all vegetation, grass, and roots to a depth of a few inches and then pile the waste in dikes around the edges.

They built sluices to let in saltwater during high tide. Salt was absorbed into the earth and then more saltwater taken in with the spring tides. The additional seawater combined with the salty soil to produce concentrated brine, which they put in oblong pots, about 200 of which were cemented together by mud in a dome-shaped kiln. The salt makers placed vents at the north and south ends of each kiln so that fire would be fanned by prevailing breezes. As the brine in the pots evaporated, workers called *malangis* added more brine, one ladle at a time, until each pot was about three-quarters full of salt crystals. The salt, which dried in piles in the open air and which the malangis then covered with reeds, was noted for its whiteness and was considered by many to be the best salt in India, yet it was also inexpensive.

This panga salt had an eager market in the neighboring provinces to the west, shipped on the River Mahanadi and its tributaries. Merchants came to Orissa to buy salt or barter with products such as cotton, opium, marijuana, and grains, carried by oxcart from central India.

Even the British in Bengal traded in Orissa salt. They needed large quantities of salt for the manufacture of munitions for their eighteenth-century wars with the French, and a significant part of the salt for their gunpowder came from Orissa.

Most of India, since ancient times, had a history of modest salt taxes. In Orissa, the Maratha, the ruling caste of much of pre-British India, levied a small tax on salt transported commercially in the province. The trade was so extensive that they could earn a substantial profit on this moderate tax and avoid a higher one that would damage the competitive price of Orissa salt. In return for this source of revenue, they looked after the promotion and prosperity of the salt trade. The Maratha rulers' attitude toward Orissa was reminiscent of a Chinese proverb: "Governing a state is like cooking small fish. It has to be done with a very light touch."

The British practiced this light touch neither in governance nor in cooking. In the late eighteenth century, Cheshire was increasing its salt production and aggressively hunting overseas markets. The empire was expected to provide these markets. Yet Liverpool salt could not compete with the price and quality of Orissa salt. In 1790, when the British requested permission to buy all the salt made in Orissa, Raghuji Bhonsla, the Maratha governor of Orissa, turned down the offer, realizing that the British were trying to eliminate Orissa salt in order to maintain British salt at an artificially high price. But when the British had their offer rejected, they simply banned Orissa salt in Bengal.

Since the border that Orissa shared with Bengal was a thick jungle, difficult to patrol, the first effect of the new ban was to create well-organized bands of salt smugglers. Inexpensive contraband salt from Orissa so flooded Bengal that the British salt still could not compete there. In 1803, in the name of fighting contraband, the British army occupied Orissa and annexed it to Bengal.

On November 1, 1804, by proclamation, Orissa salt became a British monopoly. The private sale of salt was completely prohibited. Those who had salt in their possession had to sell it to the government immediately at a fixed price. The transport of salt was forbidden. Even provisioning a ship with enough salt for the crew during a voyage had to be done under strict British supervision. Within ten years, it became illegal for salt to be manufactured by anyone other than the British government. A system of well-paid informants was established to prevent clandestine salt trading.

The earliest resistance in Orissa came from coastal chieftains, Zemindars, whose privilege and authority were undermined by the destruction of the salt industry. Before the British, the malangis in northern Orissa had been under control of the Zemindars, who had earned a good profit selling the salt made by malangis for meager wages. Workers had paid a high rent to the Zemindars for the use of coastal salt flats and manufactured salt on their own. Part of that rent had been providing for all of the salt needs of the Zemindars' households free of charge.

Still, salt workers had lived considerably better than after the British monopoly in 1804. The British advanced money to malangis against future salt production, and the malangis got deeper and deeper in debt and eventually were forced to work for the British producing salt to pay off their debt—virtual slaves to the British salt department. Thousands died every year from epidemics, especially cholera.

From the beginning, the Zemindars had obstructed British salt policy and urged the malangis, whom they controlled, to be uncooperative. The malangis began making their own salt illegally, and hundreds were arrested. In 1817, there was a rebellion in which malangis attacked saltworks and salt offices and chased away agents.

After this uprising failed, the locals gave up on open resistance but engaged in underground salt manufacture and trade. Some families supported themselves on illegal salt making.

BACK IN ENGLAND, it was well known that the Indians were angry with British salt policy. It was even mentioned in a cookbook:

One of the greatest grievances of which the poor man can complain is the want of salt. Many of the insurrections and commotions among the Hindoos, have been occasioned by the cruel and unjust monopolies of certain unworthy servants of the East India Company, who to aggrandize their own fortunes have often times bought up, on speculation, all the salt in the different ports and markets.—*Mary Eaton,* The Cook and House-keepers Complete and Universal Dictionary, *1822*

In the early nineteenth century, to make the salt tax profitable and stop the smuggling, the East India Company established customs checkpoints throughout Bengal. In 1834, a zealous commissioner of customs, G. H. Smith, was appointed, and in his twenty years in office he expanded the system into a "Customs Line" around Bengal. Salt had to pay a duty to cross this line. He was able to get taxes dropped on a series of lesser items, including tobacco, so that customs officers could concentrate on salt smuggling. Customs officers were given that always disastrous combination of broad powers and low pay. They received bounties for confiscated salt and had unchecked authority to search, seize, and arrest. Not surprisingly, bribery and other forms of corruption were widespread. In the 1840s, in its enthusiasm for enforcing this line, the East India Company constructed a fourteen-foot-high, twelve-foot-thick thorn hedge on the western side of Bengal to prevent the entry of contraband salt. After the British government took over following the 1857 "mutiny," as the uprising was labeled, the Customs Line grew until it snaked arbitrarily 2,500 miles across India from the Himalayas to Orissa. The hedge was expanded into a spiky gnarl of prickly pear, acacia, and more benign plants such as bamboo. It was impenetrable except for periodic gateways guarded by customs agents. By 1870, the Customs Line, largely dedicated to the enforcement of the salt tax, employed 12,000 people.

⁓

AT FIRST, HAVING complete control, the British wanted to produce Orissa salt and sell it in Bengal at their prices. They cleared jungle land in the coastal region to extend the salt-producing area. But British salt merchants became concerned about competition for sales in the Bengal market and lobbied Parliament to repress salt production in Orissa. In 1836, duties on domestic production were made equivalent to duties on imported salt, and from then on the government did not care if salt was local or imported because it earned the same revenue on both.

The local salt, fighting its way through a cumbersome and complicated bureaucracy, could not compete. It did not sell as fast and had to be stocked in warehouses near Calcutta and therefore risked being embezzled. The British colonial administration responded by limiting Orissa production, even closing some centers, saying that Orissa salt was of inferior quality and had a higher cost of production. In 1845, the colonial government ordered the annual production of salt to be reduced by an amount equal to half the previous year's total production.

The commissioner of Orissa, A. J. M. Mills, wrote to the colonial administration warning that reducing salt production would turn the peasants of Orissa against the British, for in the salt areas the people knew of no other economic activity.

⁓

EVEN IN THE best of times, the malangis lived hard lives in villages adjacent to salt fields. Men, women, and children—families worked the salt fields together. Some men traveled from distant villages, leaving their families behind, and lived five months of the year in temporary huts near the saltworks.

The British charged malangis for any salt lost during transportation or from inadequate storage, even though transport and storage had nothing to do with salt workers. Salt agents tried to impress on the government the need to raise payments to salt makers, but instead the government, wishing to discourage production, periodically lowered the rate.

The British policy was to preserve the jungles near the salt lands as sources of fuel wood. Since these forests had been reduced to clear land for salt production, they had an unusual concentration of tigers, bears, and leopards, and eventually the malangis were so terrified of the jungle that many refused to enter to cut fuel. In the 1846 season alone, twenty-two malangis were killed by tigers. The salt and the revenue departments both offered rewards for the heads of wild animals. Though the reward was considered substantial, it did not produce enough kills to significantly reduce the wildlife.

In 1863, the British government announced its intention to stop local salt production and instructed salt agents to end salt manufacture as soon as possible. The abandonment of salt manufacture led to a famine in Orissa in 1866. The greatest loss of life in the famine was among the malangis, because they had no crops of their own to fall back on for food. Government policy also caused a salt shortage in Bengal.

The British responded to the crisis by starting their own plant to make kartach salt. The object was to furnish locals with cheap salt while providing them with jobs. It was so successful that Liverpool salt could not compete, and so, in 1893, the government closed down the plant. Outperforming British salt was against the rules.

Once the plant was closed down, the malangis starved, while salt, their traditional cash crop, was lying at their feet in sparkling crusts, waiting to be picked up and sold. But even scraping salt off the surface of the flats was a severely punishable offense. The people of Orissa were forbidden from any activity connected with salt making. They left their starving wives and children and went to other parts of India looking for work, living in crowded, unsanitary conditions as they struggled to earn enough from menial labor to send some money to their families. In time, the malangis disappeared from Orissa, and anyone there who was poor was now deprived of salt.

⁓

THE FIRST PUBLIC meeting in India to protest salt policy took place in Orissa in February 1888, organized by the Utkal Sabha political party in Cuttack, a river port on the River Mahanadi. It was pointed out that impoverished Indians had a tax burden thirty times greater than did people in England. The tax on salt was termed "an unjust imposition of an imperial character," because the taxed salt was all imported from abroad. The government was urged to raise the income tax and save money by discontinuing the recruitment of people from abroad to the Indian civil service. These savings, the protesters at the Orissa meeting argued, would compensate the government for the loss of salt tax revenue.

In the early twentieth century, British salt policy was attacked in provincial legislatures throughout India. In 1923, to balance the budget, the government proposed doubling the salt tax. The Indian Legislative Assembly refused to support this proposal. But the British approved it anyway by decree from Viceroy Lord Reading. In 1927, the Legislative Assembly voted to halve the salt tax—

though many had called for its complete abolition. The British government did not comply.

In the Indian Legislative Assembly of 1929, Pandit Nilakantha Das, a member from Orissa, demanded the revival of salt making in Orissa and a repeal of the salt tax. The government argued that the salt tax was the only contribution to the state that poor people ever made.

The British government was not taking the issue seriously. Lord Winterlon, the undersecretary of state for India, assured the British government that there was no reason for concern about the salt issue. Not everyone in England agreed. In British Parliament, Sir Henry Craik argued that the salt tax was causing serious hardship in India and that this hardship was leading to civil unrest. Some suggested that the revenue from the salt tax was not worth the threat that unrest posed to the British Empire. Labour members warned that the salt tax could be leading them into another Irish situation in India.

In 1930, Orissa seemed near open rebellion.

And so, contrary to popular belief today, it was not an entirely original idea to focus rebellion on salt, when that idea was seized upon by an entirely original man named Mohandas Karamchand Gandhi.

GANDHI WAS BORN on October 2, 1869, at Porbandar, a small west coast town, capital of a princely state of the same name, on the Gujarat peninsula, not far from the Rann of Kutch. This is one of the reasons that, when he wanted to stage a salt rebellion, he chose this region and not Orissa on the opposite coast. Gandhi said that he felt closest to the salt makers of Gujarat.

When he was growing up in Porbandar, malangis were not a

part of his immediate world. He belonged to the Vaisya caste, the number three caste, below the ruling classes but above workers. *Gandhi* means "grocer," but Mohandas's grandfather, father, and uncle had all served as prime minister to the Prince of Porbandar. It was a small state, and its rulers exercised petty and arbitrary authority over the people while serving British rulers obsequiously. The humble house where Mohandas was born, still standing on the edge of town, testifies to the lack of wealth and position of a Porbandar prime minister. Mohandas's marriage, which was arranged when he was thirteen, lasted for the next sixty-two years. Despite his enduring reputation for living a life of simplicity and self-denial, he did not come to this easily and struggled in his youth with uncontrolled appetites, both sexual and gastronomic. In violation of his family's religious code, he experimented with meat eating, hoping it would make him large and strong like the carnivorous English.

Gandhi was a tiny man of peculiar passions and eccentric theories about sexual desire, diet, and bodily functions. Well into old age he conducted "experiments" with young women he asked to lie naked with him through the night to test his resolve to abstain from sex. He displayed a mischievous sense of humor. It is said that when asked what he thought of Western civilization, he replied, "I think it would be a great idea."

But Gandhi did not preach the superiority of Eastern culture. He said, "It would be folly to assume that an Indian Rockefeller would be better than an American Rockefeller."

He was influenced not only by his Hindu upbringing but by Jainism, which forbids the killing of any creature and whose priests wear masks over their mouths to ensure that they do not accidentally inhale an insect and kill it.

He traveled abroad, studying law in London. Visiting Paris, he gave his impression of the new Eiffel Tower: "The Tower was a good demonstration of the fact that we are all children attracted by trinkets."

In South Africa he became the leader of a movement to secure civil rights for Indians. Imprisoned for his efforts, he read Henry David Thoreau's *Civil Disobedience* in the appropriate setting— a jail cell. Along with Buddhist and Jainist writings, Thoreau was to have an enormous influence on him. He was struck by Thoreau's assertion: "The only obligation which I have a right to assume is to do at any time what I think right."

His adversaries continually underestimated him because it seemed improbable that millions would follow such an odd man. Gandhi's approach to civil disobedience was always nonviolence, but he objected to the phrase "passive resistance." It was not enough to be nonviolent. The adversary had to be opposed in such a way that he would not feel humiliated or defeated. He said that the opponent must be "weaned from error." Seeking a name for his brand of resistance, he took a suggestion from his cousin, Maganlal Gandhi: *sadagraha*—firmness in a good cause. Mohandas changed *sada* to *satya*, which means "truth." Gandhi would resist with *satyagraha*—the force of truth, a force that, he said, would lift both sides.

In all he did, Gandhi displayed an inner confidence. He was certain that his cause was right, and because it was right it would prevail. His quiet self-assurance made him a man of constant surprises—making sudden decisions and steering unexpected courses of action. When World War I broke out, this pacifist who fought British colonialism announced his support for the British war effort, thereby completely confusing his followers. Just when he appeared to be denouncing the Industrial Revolution and its machinery, he suddenly confessed his affection for Singer sewing machines. "It is one of the few useful things ever invented, and there is a romance about the device itself." Louis Fischer, his biographer who knew him personally, wrote, "A conversation with him was a voyage of discovery: he dared to go anywhere without a chart."

The other most famous Indian of his day, Nobel Prize–winning novelist Rabindranath Tagore, a tall and eloquent aristocrat, is credited with giving Gandhi his famous title, *mahatma*, the great soul, or as he put it, "the great soul in beggar's garb."

~

IN 1885, THE Indian National Congress was founded in Bombay by mostly high-caste intellectuals, including even a few Englishmen. Originally, some were even in favor of continuing British rule. But gradually they became the leading force in the independence movement. It was Gandhi who made the Indian National Congress and the cause of Indian independence a mass movement. One of the primary tools in accomplishing this metamorphosis was the *salt satyagraha*, the salt campaign.

The idea of a salt satyagraha had its beginnings in the 1929 Indian National Congress session in Lahore. While salt had become a burning issue in a few regions, it was not at the time a national issue, and despite a smoldering rebellion in Orissa and a few other places, most of Gandhi's colleagues were barely aware of it. Many in the Congress, even those closest to Gandhi, were baffled by his idea of focusing the independence movement on salt. But Gandhi argued that it was an example of British misrule that touched the lives of all castes of Indians. Everyone ate salt, he argued. Everyone, in fact, except Gandhi himself, who had renounced the eating of salt and at the time had not touched it in six years.

On March 2, 1930, Gandhi wrote to Lord Irwin, viceroy of India:

If you cannot see your way to deal with these evils and my letter makes no appeal to your heart, then on the twelfth day of
this month I shall proceed with such co-workers of the ashram
as I can take, to disregard the provisions of the salt laws. I regard this tax to be the most iniquitous of all from the poor
man's standpoint. As the independence movement is essentially for the poorest in the land, the beginning will be made with
this evil. The wonder is that we have submitted to the cruel
monopoly for so long.

The viceroy expressed his regret at Gandhi's decision to break
the law.

The ashram to which Gandhi referred was in Gujarat, across
the Sabarmati River from the city of Ahmadabad. It was an ashram
for *satyagrahis*—people committed to the force of truth—and
Gandhi had pointed out to his followers when they settled there
that they were conveniently located close to the Ahmadabad jail,
where they would be spending much of their time. The prophecy
was accurate.

Jainism was popular in the area, which meant it became a refuge
for the pests others exterminated but Jainists would not harm. It
was swarming with snakes. Gandhi lived on the ashram in a small
room the size of a prison cell. In fact, prison cells were a small ad-
justment in Gandhi's way of life. He even commented that he
could get more reading done in prison.

On March 12, 1930, Gandhi and seventy-eight selected followers left the ashram with the intention of walking 240 miles to
the sea at Dandi, where they would defy British law by scraping
up salt. A few were not from the ashram, including two Muslims

and a Christian and two men from the lowest, untouchable caste. Gandhi intended the group to be a cross section of India, but he refused to allow women marchers out of what he termed "a delicate sense of chivalry." He explained: "We want to go in for suffering, and there may be torture. If we put the women in front the Government may hesitate to inflict on us all the penalty that they might otherwise inflict."

They walked slowly along the dusty road, twelve miles a day, with the horizon rippling with heat. Marching in the lead, with a bamboo walking stick, was the bony, sixty-year-old Gandhi, his slow steps full of self-assurance and determination. Some grew too tired or their feet became too sore, and they retreated to carts. A horse was kept nearby for Gandhi, but he never used it.

The march began each day at 6:30 A.M. By then, Gandhi had been up for hours, spinning cloth, writing articles or speeches. He was seen writing letters by moonlight in the middle of the night. He stopped to speak to the villagers who gathered eagerly to see the mahatma, and he invited them to join him and to break the British salt monopoly. He also preached better sanitation and urged them to abstain from drugs and alcohol, to treat the untouchables as brothers, and to wear khaddar, the homespun cloth of India, rather than imported British textiles. In the 1760s, before the American Revolution, John Adams also had urged Americans to wear homespun instead of British imports.

"For me there is no turning back whether I am alone or joined by thousands," Gandhi wrote. But he was not alone. Along the way, local officials showed support by resigning their government posts. The Anglo-Indian press ridiculed him; The *Statesman* of Calcutta said that he could go on boiling seawater indefinitely till Dominion status was achieved. But the foreign news media was fascinated by this little man marching against the entire

British Empire, and people all over the world cheered his improbable defiance.

The press reported on his power to persuade, his determination. But the viceroy, Lord Irwin, who was being informed by British agents, was convinced that Gandhi would soon collapse. He even wrote the secretary of state for India that Gandhi's health was poor and that if he continued his daily march, he would die and "it will be a very happy solution."

On April 5, after twenty-five days of marching, Gandhi reached the sea at Dandi, not with his seventy-eight followers behind him but with thousands. Among them were elite intellectuals and the desperately poor and many women, including affluent women from the cities.

Through the night Gandhi led his followers in prayer by the warm, lapping waves of the Arabian Sea. At first light, he led a few into the water for a ceremony of purification. Then he waded out and felt his way up the beach with his spindly legs to a point where a thick crust of salt, evaporated by the sun, was cracking. He bent down and picked up a chunk of the crust and in so doing broke the British salt law.

"Hail, deliverer!" a pilgrim shouted.

ON THE OTHER side of India, the people of Orissa decided to begin salt making even before Gandhi arrived in Dandi, and they resolved to continue whether the rest of the country followed Gandhi's example or not. They opened a camp at Cuttack, to which volunteers from different parts of Orissa came and signed oaths vowing to dedicate themselves to resisting the salt laws. Regular meetings were held to discuss the nature and importance of the salt satyagraha. The British banned such meetings,

On April 6, 1930 at 8:30 A.M. *Gandhi publicly violated British salt law by picking up a piece of salt crust in Dandi on the coast of the Gujarat peninsula.* The Image Works

and people giving public addresses on the subject were arrested and imprisoned.

A public salt making was organized in Orissa on April 6 to coincide with Gandhi's. Locals blew conch shells and tossed flower petals to announce their day of nonviolent civil disobedience. As they traveled along the coast, their leader, Gopabandhu Choudhury, was arrested, but the group continued. On April 13, at 8:30 A.M., they reached their destination, Inchuri, where thousands turned out to watch them break the law.

They leaned over and scooped up handfuls of salt. The police tried to forcibly remove the salt from their hands. A crowd of dissidents ran onto the beach, picked up salt, and were taken away by the police. The protests went on for days, with waves of salt makers followed by waves of police followed by more salt makers. The police called in reinforcements. Soon the jails were filled, and more and more police and protesters were rushing into Inchuri. The police staged charges, harmless but designed to scare. It didn't work.

The salt protests spread along the coastline. A large number of the dissidents were women. Some of the salt-making demonstrations were even organized by women. The police used clubs, but the protesters remained nonviolent. After the demonstrations were over, 20,000 people turned out to throw flowers and cheer the released satyagrahis returning home from prison.

It took only a week for Gandhi's ceremonious moment on Dandi beach to become a national movement. Salt making, really salt gathering, was widespread. In keeping with Gandhi's other teachings, protesters were picketing liquor stores and burning foreign cloth. Salt was openly sold on the streets, and the police responded with violent roundups. In Karachi, the police shot and killed two young Congress activists. In Bombay, hundreds were tied with rope and dragged off to prison after the police discovered that salt was being made on the roof of the Congress headquarters.

Teachers, students, peasants—it seemed most of India was making salt. Western newspapers covered the campaign, and the world seemed to sympathize, not with the British but with the salt campaigners. White "Gandhi hats" became fashionable in America, while the mahatma remained bareheaded.

But the protest movement had spread to other groups that did not use the force of truth. One such group raided an East Bengal arsenal and killed six guards. When armored cars were sent against demonstrators in Peshawar, in the northwest, one armored car was attacked and set on fire. A second car opened fire with machine guns and killed seventy people.

Gandhi sent a letter to Lord Irwin protesting police violence, which began, as always, "Dear Friend." He then announced that he was going to march to the government-owned saltworks and take them over in the name of the people. British troops went to the village near Dandi, where the leader was sleeping under a tree, and arrested him.

The *Manchester Guardian* warned the British government that the arrest of Gandhi was a costly misstep further provoking India. The *Herald*, the official organ of the Labour Party, also opposed the arrest of Gandhi.

While India exploded, Gandhi sat in prison spinning cotton. As many as 100,000 protesters, including all of the major leaders and most of the minor local ones, were in prison. Congress committees were declared illegal. Still, the salt movement went on. The government tried to negotiate with the jailed leaders. Disapprovingly, Winston Churchill said, "The Government of India had imprisoned Gandhi and they had been sitting outside his cell door, begging him to help them out of their difficulties."

On March 5, 1931, Lord Irwin signed the Gandhi-Irwin pact, ending the salt campaign. Indians living on the coast were to be permitted to collect salt for their own use only. Political prisoners were released. A round table conference was scheduled in London to discuss British administration in India. And all civil disobedience was to be stopped. It was considered a compromise. To some, the British had won on most points, but Gandhi was pleased because he thought that for the first time England and India were talking as equal partners rather than master and subject.

Irwin suggested they drink a tea to seal the pact, and Gandhi said his tea would be water, lemon, and a pinch of salt.

Gandhi had emerged as the leading voice of Indian aspirations, and the Indian National Congress had become the primary organization of the independence movement. In 1947, India became independent, and five months later Gandhi was assassinated by a fellow Hindu who mistakenly interpreted his efforts to make peace with Muslims as part of a plan to favor them. Jawaharlal Nehru, the son of a patrician lawyer who helped found the Indian National Congress, became prime minister. Nehru was once asked how he remembered Gandhi, and he said he always thought of him as the figure with a walking stick leading the crowd onto the beach at Dandi.

⌒

IN 300 B.C., long before the British arrived, a book titled *Arthasastra* recorded that under India's first great empire, founded by Chandragupta Maurya, salt manufacture was supervised by a state official called a *lavanadhyaksa* under a system of licenses granted for fixed fees. More than a half century after the British left, salt production was still supervised by the government.

After 1947, independent India was committed to making salt available at an affordable price. Salt production in independent India was organized into small cooperatives, most of which failed. The industry is now controlled by a few powerful salt traders. The government is supposed to look after the interests of the salt workers through the Salt Commissionerate. Across the river from Gandhi's ashram, in Ahmadabad, Gujarat's Salt Commissionerate stands accused by many workers of looking after the traders rather than the workers.

The rock salt of Punjab is now in Pakistan. The west coast, Gujarat and the Rann of Kutch, has become India's major salt producer, whereas Orissa, with only six saltworks surviving into contemporary times, is no longer an important salt-producing region. Almost three-quarters of India's salt is now produced in Gujarat. Gujaratis, with their coastal economy, are not among India's poorest population. But the wages in the saltworks are so low that most salt workers come from more impoverished regions. Every year, in September, thousands of migrant workers arrive in Gujarat to work seven-day weeks until the salt season ends in the spring. They often earn little more than a dollar a day. Hundreds of workers are undeclared so that the salt traders can avoid paying them social benefits and circumvent laws forbidding child labor. Many of the workers are from the lowest caste and are hopelessly in debt to the salt producers. The glare of the salt in the dry-season sunlight renders many of the salt workers permanently color-blind. And they complain that when they die, their bodies cannot be properly cremated because they are impregnated with salt.

A storm that hit Gujarat in June 1998 decimated this cheap labor force, killing between 1,000 and 14,000 people, depending on whose count is believed. The price of Indian salt soared. But by the end of the year, the workforce had been replaced and the price had dropped back. Once again, salt could be purchased at a low, affordable price—-which every Indian citizen has a right to expect.

CHAPTER TWENTY-TWO

Not Looking Back

SOME 3,000 YEARS ago, wanting a capital in the commanding heights of the Judean hills, David conquered the fortress of Zion and built Jerusalem. Those heights have at times been a fortified high ground and at other times a peaceful gardened promenade. They offer, depending on the times, either a scenic or a strategic view of the region. Not only can a great deal of Israel be seen from here, but, on a clear day, the Moab mountains of Jordan are in view as a distant pinkish cloud. But what cannot be seen, because it is lower than the horizon—in fact, it is the lowest point on earth, 1,200 feet below sea level—is a vanishing natural wonder: the Dead Sea. The Hebrews called the sea Yam HaMelah, the Salt Sea.

About forty-five miles long and eleven miles at its widest, the Dead Sea, with the Israeli-Jordanian border running through the center of it, seems a peaceful place, of a stark and barren beauty. A first impression might be that the area is uninhabitable, and yet, like many of the world's uninhabitable corners, it has been converted, with a great deal of water and electricity, into a fast-growing and profitable resort.

The minerals in the Dead Sea give a strange buoyancy that entices tourists for brief dips. The sea is oily on the skin and doesn't feel like water. This is brine that will float more than an egg. After wading in a few feet, a human body pops to the surface, almost above the water, as though lying on an air-filled float. It is a most comfortable mattress, perfectly conforming to every part of the back—what a waterbed was supposed to be. The water, if it is water, is clear, but every swirl is visible in its syrupy density. The minerals can be felt working into the skin, and it feels as though some metamorphosis is taking place. The bather is marinating.

The Dead Sea from Picturesque Palestine, Sinai and Egypt *(Volume 2) edited by Charles Wilson, 1881.*

Pliny wrote: "The bodies of animals do not sink in it—even bulls and camels float; and from this arises the report that nothing sinks." Edward Robinson, an American professor of biblical literature, reported after his 1838 trip, with no more hyperbole than Pliny, that he could "sit, stand, lie or swim in the water without difficulty."

⌐

JERICHO, AN OASIS a few miles north of the Dead Sea near the Jordan River, which flows to the Dead Sea, is thought to be the oldest town in the world. Almost 10,000 years ago, Jericho was a center for the salt trade. In 1884, in the nearby Moab Mountains of Jordan, the Greek Orthodox Church decided to build a church at the site of a Byzantine ruin in the town of Madaba. Workers uncovered a sixth-century mosaic floor map, still on display on the floor of the church of St. George, showing the Dead Sea with two ships carrying salt, heading toward Jericho.

But the sea may have been better for transporting salt than producing it. The oily water of the Dead Sea is bitter, as though it were cursed. The area is famous for curses, the most well known being the one that destroyed Sodom and Gomorrah. The exact locations of Sodom and Gomorrah are unknown, but its residents are thought to have been salt workers and the towns are believed to have been located in the southern Dead Sea region. Since Genesis states that God annihilated all vegetation at the once-fertile spot, this barren, rocky area fits the description. But this area also has a mountain—more like a long jagged ridge—called Mount Sodom, of almost pure salt carved by the elements into gothic pinnacles.

According to the Book of Genesis, Abraham's nephew, Lot, lived in Sodom and was spared when God destroyed the town, but Lot's wife, who looked back at the destruction, was turned into salt. As columns break away from Mount Sodom, they are identified for tourists as Lot's wife. Unfortunately, they are unstable formations. The last Lot's wife collapsed several years ago, and the current one, featured in postcards and on guided tours, will go very soon, according to geologists.

In biblical times, Mount Sodom was the most valuable Dead Sea property. It was long controlled by the king of Arad, who had refused entry to Moses and his wandering Hebrews from Egypt. One of the most important trade routes in the area was from Mount Sodom to the Mediterranean—a salt route. Not far from Mount Sodom, in the motley shade of a scraggly acacia tree, are a few stone walls and the remnants of a doorway. They are the remains of a Roman fort guarding the salt route. A little two-foot-high stone dam across the wadi, the dry riverbed, after flash floods still holds water to be stored in the nearby Roman cistern.

The other source of wealth in the area besides Mount Sodom, which was mined for salt until the 1990s, is the Dead Sea. A body of water appears so unlikely in this arid wasteland cursed by God that in the afternoon, when the briny sea is a cloudy turquoise mirror reflecting pink from the Jordanian mountains, the water could easily be mistaken for a mirage.

Pliny wrote that "the Dead Sea produces only bitumen." This natural asphalt was valued for caulking ships and led the Romans to name the sea Asphaltites Lacus, Asphalt Lake. Its water is 26 percent dissolved minerals, 99 percent of which are salts. This concentration is striking when compared to the ocean's typical mineral concentration of about 3 percent.

The Judean desert, the below-sea-level continuation of the Judean hills, is a bone-white world of turrets and high walls, rising above narrow, deep canyons so pale they glow sapphire blue in the moonlight. The millions of small marine fossils embedded in the rock prove that this desert was once a sea bed, whose waters dried up in the heat that is sometimes more than 110 degrees Fahrenheit.

The seemingly barren desert hides life. There are said to be 200 varieties of flowers in the Judean desert, but they bloom only briefly and only by chance does a lucky wanderer ever see one. Graceful long-horned ibex, desert mountain goats, leap over rocks. Acacia trees, which grow in the wadis, have roots that grow as deep as 200 feet and contain salt to help them draw up the water hidden underground. A shrub known as a salt bush absorbs salt from the ground into its leaves so the leaves can soak up any moisture from the air.

The dry earth is actually laced with underground springs, some fresh, some brackish, which are easy to find because they are marked by small areas of vegetation. Each of these springs, *ein* in Hebrew and Arab, has its own history. This desert by a sea too salty to sustain life has attracted the margins of society. They have huddled along its life-giving springs: the adventurers, the pioneers, the dreamers, the fanatics, and the zealots. Many biblical references to going off "into the wilderness" allude to this area.

Across the Dead Sea, the barren, rocky, Jordanian desert is browner. The eight-mile-wide fertile strip of the Jordan Valley's east bank feeds the nation. Israel, across the river, is the land of cell phones and four-wheel-drives. Here, farmers ride on donkeys, Bedouin ride on camels, some still living in their dark wool tents, their long dark gowns elegantly furling in the desert wind.

Jordanians sometimes call the Dead Sea "Lot's Lake." Mohammed Noufal, a pleasant, graying government employee, explained that, according to the Koran, it was made the lowest point on earth to punish Lot's tribe for being homosexuals.

But if they were all homosexuals, how can they have descendants to punish?

Well, many were homosexuals.

⌒

THE CAUSE OF the Dead Sea's tremendous salinity has been the object of curiosity for centuries. In December 1100, after the knights of the first crusade made him ruler of what they termed the Latin Kingdom of Jerusalem, Baldwin I toured what is now the Israeli side of the Dead Sea. Fulcher of Chartres, who accompanied him, observed that the sea had no outlet and hypothesized that the source of its salt was minerals washing off of "a great and high salt mountain" where the Dead Sea ended to the south—Mount Sodom.

Several times in the eighteenth century, Dead Sea water samples were sent back to Europe for analysis. One such study was published by Antoine-Laurent Lavoisier, the celebrated French chemist. A number of nineteenth-century American Protestant fundamentalists tried to study the Dead Sea, some through exploration and others by way of samples. Edward Hitchcock, who taught at Amherst, concluded through samples and biblical studies that the source of the sea's salt was sulphurous springs some 125 miles away. Had he visited, he would have found brackish springs much closer.

In 1848, an American navy lieutenant, W. F. Lynch, reasoning that since the Mexican-American War was over, "there was nothing left for the Navy to perform," persuaded his superiors to finance his own expedition to the Dead Sea. Two boats with hulls of corrosion-resistant metals, in itself a considerable technical advance at that time, were specially designed and carried overland from the port of Haifa to the Jordan River. Lynch and his team sailed down the Jordan River to the Dead Sea, which he found to be "a nauseous compound." They continued sailing south for eight days and landed at the southern end of the sea. Lynch thought Mount Sodom to be not much of a mountain, which is true, and thought its composition to have a low percentage of salt, which is less true. Finding a broken-away column, he had a not particularly original thought: that it resembled Lot's wife. When he analyzed a sample, he discovered that it was almost pure sodium chloride, which either proved the salt content of the mountain or confirmed the identity of the salt pillar.

Contemporary geologists still argue conflicting theories of why the Dead Sea is so salty. According to the most widely accepted of them, 5 million years ago the Dead Sea was connected to the Mediterranean near the current port of Haifa. A geologic shift caused the Galilee Heights to push up, and these newly formed mountains cut off the Mediterranean from the Dead Sea. The Dead Sea no longer received enough water to keep up with the rate of solar evaporation, and it began to get saltier.

This theory would explain why the sea is becoming more concentrated, slowly evaporating like a huge salt pond. It is already at the density at which sodium chloride precipitates, and salt has started crystallizing out on the bottom and the edges. Bathers gingerly step over sheets of icelike salt as they enter for a Dead Sea swim.

IN THE EARLY twentieth century, Theodor Herzl, an Austrian visionary, began writing of the return of Jews to their homeland. In contemplating the viability of a Jewish state, he hypothesized that one of its valuable resources would be the Dead Sea, from which the new state could extract mineral wealth, including bromine and potash. In the 1920s, Moishe Novamentsky, a Jewish engineer from Siberia, established the Palestine Potash Company along the northern coast of the Dead Sea, in British-ruled Palestine.

In 1948, when the Arab League attacked the newly formed state of Israel, the Jordanian army crossed the Jordan River and seized most of the Dead Sea area, including the Palestine Potash Company, much of the Judean desert, and the eastern side of Jerusalem. For the next nineteen years, gun- and rocket fire crossed the border. The Potash Company was moved to Israeli-held land in the southern Dead Sea area and renamed the Dead Sea Works. The workers of the Dead Sea Works became the first Israeli pioneers in this frontier wilderness, living in rough huts with little electricity, limited water, and no air-conditioning. There too they were regu-

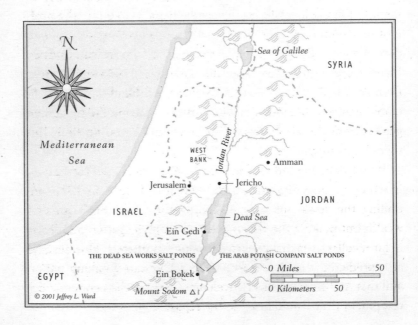

larly shelled by Palestinians on the Jordanian side. Cut off from Jerusalem and much of Israel by the Jordanians, the Dead Sea Works found its own water wells. Today the saltworks have relocated along extensive artificial ponds, and the original camp—the small houses and dining hall abandoned on a dusty plateau—is being restored as a monument to the resourceful adventurers who built Israel.

In 1956, some Israeli soldiers, having finished a tour of duty at the Dead Sea, decided to stay, drawn to a freshwater spring called Ein Gedi where a green wadi leads to a tall, thin cascade magically tumbling water out of the desert, one of only two waterfalls in Israel that flow all year. Pliny had mentioned the spot for its remarkable fertility, though, he said, it had been destroyed in war with the Romans. Possessed by Herzl's Zionist dream of making the desert bloom, the Israeli soldiers founded a collective settlement in Ein Gedi, a kibbutz, where everyone worked for the profit of the community, children were raised collectively in separate dormitories, and a paradise was to bloom on the Judean frontier. Plants were brought into a garden with lawns and tropical trees from Asia and Africa—lichees, brilliant red flamboyants, thick, climbing, green broad-leafed vines. Birds, spotting the rich green garden from the air, have made it a principal stop on their Europe-Africa migration.

It had all been predicted in Herzl's 1902 novel, *Old New Land*, in which he imagines visiting the new Jewish state in 1923 and finding the Jews not only exploiting the Dead Sea's mineral wealth but making the desert green through irrigation and living in farm collectives that exported produce to Europe. However, he also predicted that Israel would be a German-speaking nation and that Arabs would eagerly welcome Jews for the economic development they would bring to the region.

The kibbutz grew, building a health spa on the shore of the Dead Sea that offered Dead Sea mud and Dead Sea water, both of which had long been supposed to have healing properties. By the late 1960s, the kibbutz had built a hotel that today is one of Israel's leading tourist attractions.

In 1960, the Israelis built a hotel along another spring to the south, Ein Bokek. Since the Jordanians were to the north of Ein Gedi, the south was the only place for Israelis to develop. The Dead Sea Works had brought in water and electricity, and now the visitor could be offered the miracle that makes deserts livable: air-conditioning. But the Ein Bokek development was not really on the Dead Sea. Just south of the sea, the salt company had pumped Dead Sea water into a flooded area divided by dikes. There, the brine was moved from pond to pond, ever more concentrated until finally the precious salt minerals fell out of the solution in the form of a white slush that was scooped up. Still, these artificial salt ponds, concentrated to a brilliant turquoise, made a stunning sea, and sand was brought in for small beaches.

By 2000, Ein Bokek had 4,000 rooms in fourteen hotels, all with Dead Sea health spas offering a variety of treatments—an improbable oasis of white and pastel high-rises on the shores of a saltworks. The Israelis keep building ever taller hotels, and the discreet screens used by religious people to separate women's and men's nude sunbathing on the roofs of older hotels are of little help when hundreds of guests can look down from newer hotels ten stories above.

The Israeli Defense Ministry pays for every wounded Israeli veteran, and there are many, to visit a Dead Sea spa hotel two times a year. Both Danish and German government health insurance will pay for a stay in an Ein Bokek spa hotel. The Israeli tourism business has in recent years begun rethinking its markets. It has not attracted Jews in the numbers hoped for. Herzl had said that attracting the Jewish diaspora would be a slow process, but after a half century as a nation, according to the Israeli Ministry of Tourism, only 17 percent of American Jews have ever visited Israel. Christian American tourism does better, and Germans, either seeking sunshine or health spas, are the new booming trade, Israelis report without the least note of irony. Germans are the largest national group, after Israelis, to visit Ein Bokek. They also comprise one third of the visitors to Ein Gedi.

BUT THERE IS a problem.

In 1985, the Ein Gedi kibbutz built a new spa on the edge of the Dead Sea. It has the atmosphere of a public beach, packed on Saturdays with Israelis lined up like crudely formed clay statues, bizarrely coated in thick black mud, baking to a gray crust in the sun. But though the spa was originally built on the water's edge, today a trolley carries bathers to the water almost a mile away. The sea recedes from Ein Gedi about fifty yards a year.

In the first century A.D., Pliny described the now 45-by-11 mile body as being 100 miles by 75 miles. A two-lane road that used to run along the sea is now several miles inland. A flat, rocky plain that was once the sea's floor leads to the water's edge. Mountains rise up against the other side of the road, and on one rock about ten feet above the pavement the initials *PEF* are written. This is where the Palestinian Exploration Fund, a British geographic organization, marked the surface of the Dead Sea in 1917.

The greatest problem of the Dead Sea is that since the Israelis built the National Water Canal from the Sea of Galilee, the Galilee has served as Israel's primary source of freshwater, greatly reducing the flow to the Jordan River, which in turn is siphoned off by Jordanian farmers in the valley, who provide 90 percent of Jordan's produce. Not much water is left for the dying ancient sea.

Pliny called the Jordan "a pleasant stream" and said, "It progresses, seemingly with reluctance, toward the gloomy Dead Sea by which it is finally swallowed." But today the stream that approaches the Dead Sea is a little rush of silted water in a reedy gully a few feet wide. Lieutenant Lynch's specially designed boats or even a one-man rowboat could not navigate this river to the Dead Sea.

Is the Dead Sea becoming a Saharan-like sebkha, a dried bed ready for scraping? Currently, the sea loses about three feet in depth every year. Since the northern end is in places 1,200 feet deep, it is thought that the sea has several centuries left. Another theory is that it will shrink but reach a level of such concentration that it will no longer evaporate, which seems optimistic considering the ubiquitous dry salt beds in all of the world's major deserts.

A few years ago, "Dead-Med" became a popular phrase in Israel. The plan was to dig a waterway reconnecting the Dead Sea with the Mediterranean. This idea currently appears deader than the Dead Sea. The introduction of Mediterranean water would alter the composition of the Dead Sea, and mineral extraction would no longer be practical, thereby destroying one of Israel's most profitable industries.

The Dead Sea has its health spas and tourism, but the biggest business in the area, as Herzl had predicted, is the Dead Sea Works, which has even become an international company, investing in a potash mine in Spanish Catalonia, near Cardona.

The Jordanians, apparently having read their Herzl, are also counting on their Dead Sea works. The Arab Potash Company is a mirror image directly across from the Israeli company. This is the Arab-Israeli border: two sets of earthen dikes less than three feet high with a cloudy turquoise Israeli evaporation pond on one side and a cloudy turquoise Jordanian pond on the other, and in between about 100 yards of white and rust and amber soil where minerals from the two ponds leach through the earthworks.

Until a peace treaty was signed with Israel in 1996, the Jordanian Dead Sea region was a military zone, off limits to civilians. Now, at peace, Jordan has few resources but is full of plans. Mohammed Noufal observed with a smile, "All we need is Israel's technology, Egypt's workers, Turkey's water, and Saudi Arabia's oil, and I am sure we can build a paradise here."

The Jordanians too are building health spas and attracting German tourists of their own. But for them also, salt will remain the leading economic activity. There are four Israeli pumps and two Jordanian pumps moving Dead Sea water into evaporation ponds.

Sodium chloride, the salt of the past, is the first to precipitate out of concentrated brine. But hauling in bulk out of the Judean desert is too costly because of the lack of a waterway connection. The climb through the mountains is too steep for a railroad. The salt had to be hauled by truck until the Israelis built an eleven-mile-long conveyer belt. It carries 600 to 800 tons of salt in seventy minutes to the town of Tzefa, where the land is then level enough for a railroad to the Mediterranean.

This system is still too expensive for sodium chloride to be profitable. But the Israeli company sells 10 percent of the potassium chloride—potash—in the world, a product much in demand for fertilizers. It also produces liquid chloride and bromide for textiles and pharmaceuticals and methyl bromide, a pesticide. Under pressure because of damage to the ozone layer, it is phasing out methyl bromide production. The Jordanians say they are thinking of starting it up.

The Dead Sea Works believes its future is in magnesium. Magnesium chloride, what Lieutenant W. F. Lynch called "a nauseous compound," is the salt that gives the sea its bitter unpleasant taste. It is a slightly more expensive but less corrosive alternative to sodium chloride for deicing roads. From magnesium chloride, the Dead Sea Works also produces magnesium, a metal that is seven times stronger than steel and lighter than aluminum. The company has invested in a joint venture with Volkswagen to make car parts. Will one more Herzl prediction come true and Israel become German-speaking after all?

Once the sodium chloride precipitates out, falling to the bottom of the pond, the principal target mineral is $6H_2O$ $MgCl_2$ KCl. This grayish crystal sludge, called carnolite, fuses potassium chloride, sodium chloride, and magnesium chloride into a single crystal.

The sodium chloride that precipitates out before carnolite is allowed to fall to the bottom of the pond, constantly raising the height of the pond bottoms. The company keeps building the dikes higher, but the raised ponds have been flooding hotel basements, to the great irritation of the tourism industry. The Dead Sea Works counters that its workers were the pioneers who dug the wells and provided the water and electricity that made the area usable in the first place. Tensions persist. This is, after all, the Middle East. The Dead Sea Works, recognizing the problem, has started a flood prevention program to help hotels.

Common salt has become a nuisance.

CHAPTER TWENTY-THREE

The Last Salt Days of Zigong

I N FEBRUARY 1912, ancient China came to an end when the last of three millennia of Chinese emperors abdicated.

Imagine twentieth-century Italy coming to terms with the fall of the Roman Empire or Egypt with the last pharoah abdicating in 1912. For China, the following century has been a period of transition—dramatic change and perpetual reevaluation.

After 1912, the new Chinese republic struggled economically at a time when World War I was consuming the treasuries of Europe, blocking loans that might otherwise have been available for a young and embattled government. With Western encouragement, the Chinese went back to one of the old ideas of the emperors. Salt could fill their treasury.

In April 1913, the new Chinese government obtained a Western loan from the Quintuple Group of Bankers of £25 million. The entire revenue of the salt administration was put up as security to repay the loan. The salt administration that the republic inherited from the emperors was elaborate but extremely corrupt. In order for it to regain credibility in the eyes of Western bankers, the Chinese had to place a foreigner in charge of purging the system. An Irishman named Sir Richard Henry Dane was hired to be chief foreign inspector to the Chinese government. Amusingly, in the hindsight of history, Dane's great qualification for this post was that he had been inspector general of the Salt Excise in India at a time when Indian salt policy was still considered an enormous success. Dane himself freely admitted that in India local salt had not been able to compete with Liverpool salt, but what was viewed as the important accomplishment in India was that the administration had derived enormous revenue from salt.

Nicknamed "the salt king," Dane seems to have been a British colonial cliché, complete with bushy mustache, walking stick, and a reputation for big-game hunting. According to a 1917 issue of the American magazine *Asia*, when the Chinese offered him the position, he had been about to embark on a two-year "hunting trip in the wilds of Africa."

Interviewing the new republic's new salt king, *Asia* magazine described Dane as "shaggy and blustering."

"I suppose you Americans know nothing about Chinese salt or its administration," he began a bit testily, as he finished a scone and set down a cup of tea in his comfortable apartment in Peking.

In the old administration, salt was taxed along the road from producer to consumer. To cross Hubei Province, one had to pay forty-two different taxes. In theory, salt production was a government monopoly, but in practice China was too large for the government to control all salt production, trade, and transport. Instead, it merely tried to control commerce, by authorizing an elite group of merchants to transport the salt from place of production and then taxing the transportation. These elite firms, known as Yuen Shang, were usually family owned and either rented out these rights or maintained them as a family monopoly from generation to generation. In Chinese folk literature, the salt smuggler is always a hero fighting the evil and corrupt salt administration. The villain of the story is often not the government but the Yuen Shang.

Salt merchants amassed great wealth and liked to display it. Both Shaanxi Province, north of Sichuan, and Shanxi Province, north of that and near Beijing, are famous for elegant mansions built by seventeenth-century salt merchants. In Suzhou, a city of canals some fifty miles west of Shanghai, best known for its silk merchants, the gardens that have become one of China's leading tourist attractions were built by salt merchants.

Smuggling was widespread. Dane was informed that half the salt consumed in China at the time was smuggled. The Yuen Shang took advantage of the lack of a standard unit of measure to carry more salt than they reported and sell the surplus on the black market. Boatmen and cart drivers were able to bribe inspectors and make profits from smuggling. Dane estimated that 40,000 people were engaged in mostly illegal salt traffic on the Yangtze River alone, involving many thousands of square-sailed salt junks. He organized the Salt Preventive Service with salt police stations at strategic points, but this failed to stop the smuggling.

In *Strange Tales from a Chinese Studio*, Herbert A. Giles described the smuggling he observed on a trip from Swatow to Canton in 1877:

Apropos of salt, we came across a good sized bunker of it when stowing away our things in the space below the deck. The boatman could not resist the temptation of doing a little smuggling on the way up. At a secluded point in a bamboo shaded bend of the river, they ran the boat alongside the bank, and were instantly met by a number of suspicious-looking gentlemen with baskets, who soon relieved them of the smuggled salt and separated in different directions.

⌒

DANE ASSERTED THAT "it is the salt revenue that has been safeguarding the credit of China. . . . Salt has always formed one of the principal sources of government revenue but since June 1913, when the reform administration was inaugurated, it has leaped into first place." Until 1915, maritime custom was the leading government revenue source. But Dane claimed that once he had reestablished a centralized salt administration in 1915, salt revenue increased over the previous year by 100 percent.

Dane found the Chinese to be heavy salt consumers, notably higher than in India. He asserted that the Japanese were the heaviest salt consumers in the world and that the Chinese consumed at about the same rate, which he estimated to be about twenty pounds per capita.

It is probable that neither Chinese nor Japanese consumption was as high as American, but the fact that Japan would even have such a reputation was remarkable considering its unsuitability for salt production. Japan has a long and meandering coastline that would otherwise provide ideal tidal ponds and inlets for sea salt production, but its humid climate with regular storms and periodic flooding renders it a salt region of high cost and low production.

Historically, the Japanese depended on imported salt, but in the late nineteenth and early twentieth centuries a modernization plan under Emperor Meiji built a strong centralized economy and a modernized military. Newly empowered Japan thought it was unwise to be dependent on foreign salt. In January 1905, the Salt Monopoly Law came into force, establishing twenty-two salt offices around the country to regulate production, which became a state monopoly. The Japan Monopoly Office set prices and ended imports.

Salt production was concentrated on the Seto Inland Sea, which, though far from ideal, was deemed the best sea salt climate in Japan because it is sheltered between two islands in a relatively southern climate. The main area, from Osaka to Hiroshima, was devastated in World War II, but the beds were restored in the 1950s. Industrialized Japan has remained self-sufficient in edible salt for products including pickles, salted fish, soy sauce, and miso. Miso is a Japanese offshoot of the Chinese import soy sauce, and like soy sauce, it is made from salt-fermented beans.

Traditionally, though less so today, a Japanese meal ended with pickles, and in the north pickles are served with afternoon tea. Japanese homes almost always had a smell of pickling, which is one reason most Japanese now prefer to buy their pickles. Among their favorite pickles are eggplant, Chinese cabbage, radish greens, and mustard greens, which are added to rice. Daikon, a root that is curiously known as either a Japanese radish or a Chinese turnip, is a staple of Buddhist monastaries, pickled in alternate layers of salt and rice bran.

Dane found that the Chinese also "use a great deal of salt for soaking and preserving vegetables, salting fish, pickling and preserving meat." This was why the Chinese and Japanese were heavy consumers of salt.

AT THE TIME Dane went to China, as throughout history, most Chinese salt was sea salt, pumped into evaporation ponds by windmills. However, Dane said, "The best salt in China is that produced from the salt wells of Sichuan." Sichuan produced about one-fifth of China's salt.

Dane had arrived at the end of a golden age of Sichuan salt, that had begun in the eighteenth century. The salt wells were mostly located around what became the city of Zigong. Between 1850 and 1877, there were 1,700 salt merchants in Zigong, and 20 percent of salt production was held by four families that had accumulated fabled wealth.

Zigong grew along a curve of the Fuxi River, a gracefully winding tributary of the Yangtze, clogged with shallow, flat-bottomed, oar-powered boats that carried salt to much of central China.

The Yangtze, the 3,700-mile waterway from the Tibetan mountains to the port of Shanghai, the third-longest river in the world, divides China into its north and south, and yet, until the 1949 Communist victory, China had so little transportation infrastructure that there was not a single bridge crossing it. The Yangtze was the key transportation artery through China and its tributaries, the only connection between north and south China.

By the seventeenth and eighteenth centuries, salt merchants were traveling regularly to the little provincial town of Zigong. In 1736, merchants from Shaanxi Province began construction of a guild hall in Zigong for out-of-province salt traders. It took sixteen years to complete this palace with roofs fanning out like wings in all directions, nimble stone dragons leaping from the edges. The courtyards were lined with red pillars, not the usual wooden pillars, but stone ones painted red.

Long before the red star and red flag of Communism found their chromatically perfect home in China, red was the Chinese color. The symbol of happiness, it was the color worn by a bride at her wedding. And so the salt merchants built a red palace with gilded carvings depicting Daoist legends. The guild hall, like many Chinese houses of the period, used no 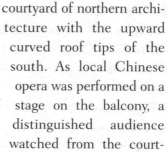 nails but was held together by fitted joints. It combined the four-sided courtyard of northern architecture with the upward curved roof tips of the south. As local Chinese opera was performed on a stage on the balcony, a distinguished audience watched from the courtyard that was gardened with tall trees and elegant dwarf bonsais.

Jealous of the out-of-town merchants' showy guild hall, the local salt merchants in Zigong built their own red-pillared, wing-roofed mansion, a temple with a commanding view of commerce on a high bank over the curving Fuxi, from where they could view the congested traffic of flat-bedded boats rowing cargoes of salt to the Yangtze.

The well technology that had been ahead of the world in the Middle Ages continued to improve. One advance was the addition of four oxen driven in a circle, attached to a pole, which wound and released tough rope braided from bamboo leaves. The rope system was counterweighted with huge rock slabs and ran to a large wheel that served as a pulley and then to the top of the derrick to control the bamboo tube that was dropped down for brine. The longer the tube, the higher the wooden derrick that raised and lowered it.

The brine was piped to gas-heated pans in the boiling house.

Then a ladle of ground yellow bean, soya, and water would be added. After about ten minutes a yellow scum would form on the surface and be skimmed off, ridding the salt of impurities with a simpler formula than Europeans ever found. After the brine had been boiled five or six hours to pure crystal, the salt was shoveled into a barrel and hardened.

In 1835, a new well, the Shen Hai well, was drilled in Zigong. At 2,700 feet, it struck natural gas. At 2,970 feet, the well reached natural brine, but the drilling continued down to 3,300 feet, making it at the time the deepest drilled well in the world. Twenty-four years later, an American would be cheered for the achievement of having drilled 69.5 feet in Titusville, Pennsylvania.

The Chinese used oxen until 1902, shortly before Dane's arrival, when coal-fired steam engines were introduced. In the nineteenth century, the Zigong ox herd was usually about 100,000 head. Because of the oxen, in Zigong, unlike most of China, beef, albeit very tough beef, was part of the working-class diet. At the rig where they labored, salt workers would boil the tough old ox meat until it was tender, and then they would add the most common Sichuan seasoning, *ma-la*.

Unique to Sichuan, *ma* is the spicy flavor of a wild tree peppercorn called *huajiao*—with a taste between peppercorn, caraway, and clove, but so strong that too much will numb the mouth. Two varieties grow in Sichuan, clay red peppercorns and the more perfumey brown ones. *La* means "hot spice" and is accomplished with small burning red peppers. The combined seasoning, ma-la, defines the taste of Sichuan food.

Another specialty of Zigong salt workers was *huobianzi*. The tough thigh of an old salt well ox was cut by hand in a continuous paper-thin slice by slowly turning the leg. Some pieces could be two yards long. Zhang Jianxin, managing director of

Sichuan Zigong Tongxin Food Corporation in Zigong, where huobianzi is still made today, complained that it is difficult to get a leg as tough as the legs from the old working oxen, but some farm animals too old to work are satisfactory.

The strips were seasoned with soy sauce and salt, then air dried and grilled over a low heat from burning ox dung. Today, a gas heater is used, but it is said that huobianzi that is cured over ox dung has "a special fragrance." It is served with a vegetable oil containing hot peppers.

Meanwhile, the wealthy salt merchants went for more exotic fare. In China, the more obscure the ingredients and the more arcane the method, the more status the dish has. "Soaked frog" was a specialty for Zigong salt merchants. A few pieces of wood would be floated in a large jar of brine. Live frogs would be put in the jar, and they would desperately perch on the pieces of wood. The jar was closed and sealed. After six months, the jar would be opened, and the frogs would be dead and dried on the wood but preserved because they had dipped in the salt. They would then be steamed.

The salt merchants were also fond of stir-fried frog stomachs. Unfortunately, a frog's stomach, however tasty it might be, does not go a long way. It is said in Zigong that to get one serving of fried stomachs, a cook would kill 1,000 frogs.

⌒

THE CHINESE CONTINUED percussion drilling in Zigong even after the American oil industry had developed much faster techniques. Their homegrown technology was slow but reached depths that, even in the age of petroleum, were astounding. In the 1920s, the Chinese drilled a well to 4,125 feet, and in 1966, the Shen Hai well, a record breaker in 1835, was drilled even deeper to 4,400 feet, about four-fifths of a mile.

The Chinese character for *jing*, meaning "well," is a depiction of a Zigong derrick. The derricks, towers of gray, weather-beaten tree trunks lashed together high in the air, rigged with bamboo leaf ropes, dotted the Zigong landscape the way oil wells do in petroleum cities.

In 1892, Sichuan salt makers discovered the layer of rock salt that feeds the groundwater under Zigong. Today, Zigong produces more rock salt than brine salt. But in the first few decades of the twentieth century, between 300 and 400 brine wells were operating in Zigong.

The beginning of the end for the ancient Sichuan salt industry came belatedly in 1943, when for the first time a rotary drill, bore a well in Sichuan. It took another twenty years for the change to become apparent. In 1960, Zigong was still a backward provincial town of a third of a million people living among medieval brine derricks. That year, the last percussion-drilled shaft in Sichuan was completed. Along with modern rotary drills and rock salt mining, Sichuan salt producers were soon using vacuum evaporators, making modern white salt with crystals of a uniform size.

It was in the 1960s that Zigong got its first "modern" public transportation. As brine boiling was fading, Sichuan engineers

The character jing, *on Derrick brand soy sauce labels.*

Brine well derricks in Zigong in the early 1960s. Photo by Yu Minyuan.
Zigong Salt History Museum

found a new use for the natural gas at the wells. Buses were built with giant gray bladders on the roofs, filled with the local natural gas. They started out on their routes with the huge rectangular bladder on top almost as big as the bus. The big bladder swayed and jiggled like Jell-O as the bus rounded corners, and then it gradually deflated, the gray bag sagging from the roof, as the gas was used up. Locals call the buses *da qi bao*, which means "big bag of gas." The buses need frequent refueling. Today, with Zigong tripled in population, the old buses are considered an embarrassing eyesore, and the remaining ones are left with the undesirable rural routes.

ZIGONG IS NOW a sprawling city of 1 million people, including residents of the suburbs. Stone-edged holes in the ground are all that remains of many wells. Only a few derricks are left standing in the hilly municipality, though many were not torn down until the 1990s, some as recently as 1998. Scholars struggle quixotically to save them, but these are not good times for landmark preservation in a China passionate about modernization. In 1993, two twin derricks, the symbol of Zigong, one 290 feet high and the other 284 feet high, were torn down. They were dangerously decrepit, and the government would not spend the money to repair them. "They didn't understand the value, that these things are only in Zigong," said Song Liangxi, a Zigong historian.

The Shen Hai well, the rugged old contraption of tree trunks and rocks, still operates. As with hundreds of wells that once pumped in Zigong, the threshold to the front gate is two feet high—to symbolically keep the wealth inside. The well has ten workers, who keep it operating twenty-four hours a day. A cable slowly lowers into the earth for several minutes and then emerges with a long wet bamboo tube that is held over a tub by a worker who pokes the leather valve at the bottom of the tube, releasing several bucket-loads of brine. The brine is still evaporated in pans heated by the gas from the well. In 1835, when the well was drilled, it had an estimated 8,500 cubic meters of gas. In the year 2000, the operators believed it had 1,000 cubic meters left.

The Shaanxi guild hall remained a guild hall until the fall of the last emperor. Then it became a local headquarters for the Chinese nationalist movement of Chiang Kai-shek. After the Communists came to power, Deng Xiaoping, a native of Sichuan who became secretary general of the Chinese Communist Party, decided to make it a salt museum.

Today in Zigong, there are still some crumbling tile-roofed

Chinese houses with the roof tips turned up in the southern style, but most of them are in disrepair, seemingly awaiting demolition. The new buildings seem kitschy spoofs on urban high-rise architecture. As in Beijing, historic monuments were torn down to make way for buildings that will never be completed, that remain concrete and exposed steel rods because the companies building them went bankrupt. But the guild hall is preserved as a national monument.

Of greater interest to the locals than the guild hall is the small amount of salt still made at the Shen Hai well. They call it flat pan salt and believe it is better for pickling than the industrial salt made in vacuum evaporators. Paocai and zhacai makers want flat pan salt for their pickled vegetables. It is sold in the Zigong market, but outside of Zigong, this medium-grained, untreated salt is becoming difficult to find. Zhang Jianxin at the Sichuan Zigong Tongxin Food Corporation wants flat pan salt for his huobianzi and other products such as *larou*, a traditional Sichuan cured pork. Zhang Jianxin's recipe for larou is as follows:

Cut pork into pieces any size. Cover with salt and spices including huajiao, leave it a week, wash off the salt, hang it four feet above a charcoal fire and smoke it slowly for two days. Add peanut shells and sugarcane trimmings to the charcoal. People at home add cypress leaves.

But Zhang Jianxin has trouble finding a salt that he wants to use. "Vacuum salt is too fine-grained and also they add chemical things I don't like," he said. The added chemical to which he referred is iodine, which he said has a taste that "is bad for our product."

SICHUAN PROVINCE IS the size of France with twice the population. In the mid–twentieth century, when the Chinese population expanded at an unprecedented rate, the number of inhabitants in Sichuan grew to its current 100 million people, most of them crowded into the eastern half of the province. The west is a desert leading to Tibet. Sichuan also has a bamboo mountain forest that is home to the earth's only remaining wild panda population. But most of the province is subtropical, like the American South.

The Sichuan landscape is a tribute to the water management skills of the heirs of Li Bing, the third century B.C. governor, with dikes and sluices breaking up waterways into a lush green quilt of flooded rice paddies, dark-soiled vegetable patches, cypress groves, and bamboo stands. Soil erosion is rare and wasted space even rarer. But despite this rich agriculture, the farmers seem poor. They produce an enormous quantity of food, but in their villages built along the dirt trails that connect paddies and fields, there are too many people. They live in patched and crumbling mud-and-straw houses, a few still decorated with huge posters of Mao.

Children hike miles to school along the dikes between rice paddies and up into the green mountains. Women with brightly colored parasols carry children on their backs in wicker strap-on seats that are made only in Sichuan. A frequent sight in the Sichuan countryside, one seen in Marco Polo's China, is noodles more than seven feet long, hung out to dry like laundry on a line.

Although most of the big derricks have been torn down, a few small brine wells remain. One in Dayin, west of the Sichuan capital, Chengdu, had a single post the height of a telephone pole. The well was only 1,000 feet deep, a considerable depth by any but Chinese standards, but at that relatively shallow depth the brine was weak, only 10 percent salt. That was why the Chinese learned deep drilling.

A farmer in a worn blue Mao jacket who grew grains, vegetables, and sweet potatoes on the land said that in the 1960s salt had been his most profitable crop. Asked who built the well, he shrugged and said, "Oh, that well has been there forever."

The blue jackets, which during the 1950s and 1960s were the only clothes available in China, along with matching pants and caps, are still commonly seen in the countryside. It is not a political statement, just people too poor for new clothes. The match-

The Dayin well.

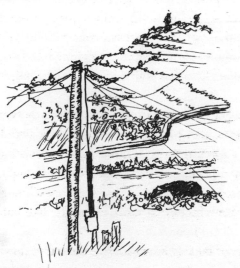

ing pants are seldom seen anymore. Pants wear out, but a good jacket lasts forever.

Next to the pole at the Dayin well, there was a stone stool. A lone farmer would sit on the stone with his feet peddling a bamboo wheel, which would raise and lower a bamboo tube into the 1,000-foot hole. The brine was piped into a tank, over which stood a much larger bamboo wheel about ten feet tall, with bamboo cups lashed to its rim. This larger wheel was turned by a man walking carefully inside the wheel, a simpler version of the medieval wheel in Salsomaggiore. The wheel would scoop up the brine and drop it on top of a wall of dried branches. As the brine dribbled down the branches, with the help of wind and sun, it would become more concentrated. After it dripped into the tank below, it was ready to be boiled for evaporation. Since this well had no natural gas, coal, which is abundant in the area, was used for fuel.

In 1998, the government salt corporation sealed the well, capped the small hole in the ground with concrete, along with many other wells in the area, and ruled that such salt was substandard and illegal to sell.

"But there's still brine there," the farmer insisted.

By the standards of Chinese history, salt producers are no longer tightly controlled. The tax is on selling, not producing, and it is no longer a major source of revenue. But the iodine requirement, the reason the little well in Dayin was capped, is often seen as a new form of government control of salt.

The World Health Organization and UNICEF urge salt producers to include iodine in their salt to prevent goiter, an enlargement of the thyroid gland. Since everyone uses salt, it is an ideal distribution vehicle. They claim that 1 billion people worldwide are at risk of iodine deficiency. In addition to thyroid enlargement, symptoms of iodine deficiency can include nervousness, increased and irregular heart rate, and muscle weakness. Iodine deficiency can also lead to mental disability in children.

Iodine was used to cure goiter even before it was known to be iodine. Humphry Davy, among others, had suspected that iodine was an element, but it was Jean-Baptiste Dumas, the French chemist and founder of one of the first schools of industry in France, who, in 1819, proved that iodine was present in natural sponge, which had been a standard treatment for goiter.

In treating goiter, once again, China was centuries ahead of the West. A fourth-century-A.D. Chinese physician, Ko Hung, prescribed an alcoholic extract from seaweed for goiter. Many seaweeds are rich in iodine, which is why the Japanese, who not only eat a great deal of seaweed but fertilize crops with it, have had relatively little experience with the disease. In China, as in most of Asia, goiter has little history in coastal regions but has often been problematic in mountainous interior provinces, including Sichuan.

American salt is usually iodized. The British, having few cases of goiter, do not iodize, and the French sometimes, but not always, iodize their salt. Among afflicted populations, iodized salt is well appreciated. Myanmar, formerly Burma, has an iodized salt policy, but the tribesmen in the remote highlands cannot get the treated salt and instead trade illegally across the Chinese border for it. In exchange for Chinese salt, which they believe will help with their goiter problem, they offer rare, endangered wildlife species. The Chinese value these animals for folk medicine. The tongues of the antelopelike serow are thought to cure headaches, and the nimble legs of goatlike gorals are ground into a powder used on aching joints. Rare Himalayan black bears are killed for their gall bladders, which are used to treat liver and stomach ailments. The commerce across the Myanmar border is especially tragic because much of this black-market Chinese salt is in fact not iodized and so will not help them with their goiter problem.

Iodized salt has become controversial in developing countries where government control of salt is a historic issue. In 1998, India followed China's 1995 decision and, under pressure from the world health community, banned the sale of noniodized salt. In both countries, the move was popular with health authorities, doctors, and scientists, but very unpopular with small independent salt producers.

As China became a modern state, its salt became modern salt—small uniform grains with iodine added. And like other modern people, the Chinese have started longing for salt that is a bit more irregular, perhaps less pure. Impurities are things that were left in, and many prefer this to chemicals that are added. The controversy over iodized salt is in part the distrust of chemical additives that have become part of life in virtually all cultures. In the Jewish religion, most rabbis state that salt must be noniodized to be considered kosher for Passover.

In Sichuan, wary consumers insist that iodine gives salt a peculiar taste. But small producers also suspect that the ban is a government conspiracy to put them out of business and once again give state salt companies a monopoly. Peasants, such as the family at the little foot-operated well that was capped in rural Dayin, do not have the knowledge or money to meet government standards for iodized salt.

In September 2000, the Indian government repealed its ban on noniodized salt under pressure from Hindu nationalists and Gandhians who recalled Gandhi's assertion that every Indian had a right to make salt. But the Chinese authority did not seem in-

clined to go back on its decision to ban noniodized salt. Li Fude
of the government salt agency for Sichuan Province, the General
Sichuan Salt Company, said, "It was decreed by the prime minister himself." He said it like an ancient bureaucrat speaking of the
emperor.

CHAPTER TWENTY-FOUR

Ma, La, and Mao

THE CHINESE HAVE been slow to part not only with their emperors but with many ancient ideas. Among the lingering old ways in modern China are attitudes about food—about salt and seasoning and how to construct a meal. Many of these ideas, though notably different from current Western thought, did exist in the pre-Rennaissance West. The differences between China and the West on food are far greater today than 1,000 years ago.

The Chinese seem ready to eat anywhere and anytime. City streets and rural roads are lined with food stalls. On the trans-Siberian railroad that runs from Moscow to Beijing, the heater at the end of cars that the Russians use to make tea is used by the Chinese to prepare whole meals. They crowd into the dark, closetlike space and chop vegetables and spread out seasonings. Not only do they cook and eat constantly, but they talk a great deal about the meaning of their foods. Food sometimes seems a

Chinese obsession, and the culture at times seems almost afflicted with epicurism.

The contemporary Chinese novelist Lu Wenfu wrote:

> The word gourmet is pleasing to the ear, perhaps also to the eye. If you explain it in simple everyday language, however, it's not so appealing: A gourmet is a person who is totally devoted to eating.—Lu Wenfu, *The Gourmet*, 1979

In China, southern food, especially Cantonese, is usually said to be the best. But after 1949, when Mao Zedong from Hunan and Deng Xiaoping from Sichuan came to power, the hot spicy food, la, from southwestern China, came into official fashion. "If you don't eat la, you are not a revolutionary" became a popular saying.

In 1959, a restaurant for the political elite was established in a Beijing house of gardened courtyards built for the son of a seventeenth-century emperor. Predictably, it was a Sichuan restaurant, and was simply named the Sichuan Restaurant. Zhou Enlai, the long-time premier, and Deng Xiaoping were regulars. For years it was considered one of the few good restaurants in Communist Beijing.

The restaurant remained a symbol of the times when, in 1996, its antique setting was bought by a Hong Kong entrepreneur, who turned the house into a private members-only club with the obsequious gentlemanly service reminiscent of British colonialism. The so-called Chinese takeover of Hong Kong has in fact meant that many Communist Party relics have been bought up by Hong Kong entrepreneurs. The Sichuan Restaurant has survived in three less sumptuous Beijing locations. Its head chef, Yu Jiamin, is a native of Beijing who, in 1970, at age nineteen, began apprenticing in Sichuan cuisine. "For me, it is the most complete cuisine, the only one that completely uses six flavors," he said.

The notion of balancing principal flavors is central to Chinese cooking. The six of Sichuan food are expressed as a musical jingle: "ma, la, tian, suan, xian, ku." Ma, the spicy huajiao, is the sixth flavor unique to Sichuan, though la, hot peppers, is also typical of the area. *Tian*, meaning "sweet," *suan*, meaning "sour," *xian*, meaning "salty," and *ku*, meaning "bitter," are universal.

Each dish will have a flavor, or ideally a combination of flavors, ma-la being the most famous Sichuan combination. Xian, salty, is the most used flavor, a central motif. It is considered a balance to all the others. Salt is believed to bring out sweetness and moderate sourness. In ancient times, tea was prepared in Sichuan with salt and ginger added. Salty and spicy, xian-la, is such a popular Sichuan combination that it has been bottled in the form of soy sauce and hot peppers. Xian-la is also a recurrent theme in other warm climates from Cajun Louisiana to Vietnam, where ground hot pepper and salt are served on limes, grapefruit, or pineapple to moderate the acid taste.

In China, meals are put together by counterbalancing these combinations. Balance, making a complete flavor by blending opposites, like combining an acid and a base in chemistry, is an ancient concept in cooking. The fourth-century-B.C. Chinese belief that the world is made up of two opposing forces, yin and yang, has long been applied to cooking. The Chinese classify foods into warm and cold according to their attributes, not their temperature, similar to the way Europeans classified and balanced foods in the Middle Ages. All cooks do not agree on which foods are hot and which cold, but fat meat, hot spices, and alcohol are usually thought to be hot, while bland vegetables and fruit are usually considered cold. In the West, such ideas trace back at least to

Hippocrates in fifth-century-B.C. Greece. Some scholars believe the idea originated in Greece and spread to Asia through India. Others argue that different cultures thought of it independently. Some scholars believe that indigenous North Americans held these beliefs before the arrival of Europeans. Such ideas were the basis of the Church's lean and fat day interdictions. But in time they degenerated in Europe to such frivolousness as Grimod de La Reynière's distinction between blond and brunette food—as with women, he preferred his food blond.

Ancient concepts such as hot and cold foods are still seriously discussed in China. Dishes that are ma-la or dishes that are very salty are contrasted with bland dishes. But also tian, sweet, is considered a good counterbalance to ma-la. *Tian shao bai*, which literally means "sweet white stew," consists of thick bacon strips stuffed with sweet bean paste on a bed of sweet rice and sprinkled with sugar. Like many Chinese dishes, this sounds repugnant by itself. But a bite of tian shao bai is a perfect moment, almost an antidote, when the mouth is aflame from a bite of a ma-la dish.

This idea of using sweet as a countermeasure to salty or spicy used to be common in the West. Apicius prescribed adding honey to a dish that is too salty. Pliny phrased it in reverse: "Salt corrects our aversion when we find something over-sweet." In medieval Catalonia, salt cod was served with honey. Platina prescribed sweetness as a counterbalance to la: "Sugar softens and tempers all dishes of hot and aromatic spices." This is the reason the people of Collioure make their spicy sweet Banyuls wine to accompany their salty anchovies. But in the eighteenth century, *dessert,* a word from the French verb meaning "to clear the plates," became such an elaborate showpiece in Europe that sweet was gradually eliminated from the rest of the meal.

When the dessert idea was first taking hold, a dessert was sometimes served at the end of each "course," and a course was often a combination of dishes. In China, a course is still an assortment of foods in the middle of the table, often on a large rotating disk, a lazy Susan, which makes all the platters easily accessible. People sit around the table with only a small plate or bowl and with chopsticks take a bite of one then another dish, mixing the combinations—a bite or two of hot, then a taste of sweet.

In all of the courses, vegetables play a significant role. In Sichuan, wild mountain vegetables such as mushrooms are a specialty, as are numerous varieties of bamboo shoots eaten raw, cooked, or preserved in salt.

The first course is usually an assortment of foods that are cold in temperature, the second an assortment of heated ones. The last course, especially in Sichuan, where by this point the palate has been through a great deal, is bland, usually a very bland soup. Sometimes a course of white rice is served before the soup with very salty paocai. Rice is usually not served with the other courses. Except among the very poor, many meals do not include rice at all.

HISTORIANS DEBATE EXACTLY why food in China is seasoned with products fermented or pickled in salt, and not with grains of salt added directly to food. The idea of producing saltiness without the direct use of salt is Asian, though it is not that different from the Roman use of garum. The following recipe for the Sichuan classic *huiguorou* is an example of cooking with salted condiments—in this case three—but without using salt directly. The recipe is by Huang Wengen, a cooking instructor at the only accredited school of cooking in China, which is in Chengdu, the capital of Sichuan.

For authentic huiguorou you must have these ingredients:
pork thigh, garlic greens, douban, dousi, soy sauce, sugar, and msg.

Boil the ham until it is nearly cooked. Cool it. Cut thin slices perpendicular to the bone.

Chop garlic greens.

In a wok with mixed vegetable oil:

Stir-fry meat until the slices begin to curl a little.

Add douban and dousi. When the sauce turns reddish add soy sauce and a pinch of sugar and a pinch of msg (use small amounts of all these ingredients). Finish with the chopped garlic greens.

Like so many Chinese dishes, this one uses pork. The eighteenth-century French naturalist Georges-Louis Leclerc de Buffon speculated that the reason Islamic proselytizing was not very successful in China was the Islamic rejection of pork. The Chinese not only cook fresh pork but also have a long tradition of salt-curing pork into bacons, hams, and sausages. In 1985, the pig population of China was estimated to be 331 million, which is far greater than that of any other country in the world. According to a survey of rural China conducted from 1929 to 1933, pork and pig lard accounted for 70 percent of animal calories consumed. The cooking oil, usually a blend of sesame, peanut, and other vegetable oils, that is used in so much Chinese cooking is often a modern, healthier substitute for pork fat.

According to Huang Wengen, "You cannot cook Sichuan food without douban. We went to France, and I brought douban because the douban in France is no good. It was a six-week cultural exchange program of cooking teachers—a school near Lille by the narrowest part of the Channel, Le Touquet. But it was impossible to teach Sichuan without the products. Huajiao, for example. We brought what we needed that was practical to carry: huajiao, douban, dousi, zhacai."

All of these irreplaceable ingredients except huajiao are salt products. Zhacai is vegetables in salt. Douban is a bean paste from a big, flat, green soybean that is dried until it turns hard and yellow and is then fermented with salt and hot pepper. Dousi is a black paste made from fermented yellow beans, very salted but without chili.

Another ingredient seen by the Chinese as a salt alternative is MSG, or monosodium glutamate. While it has no flavor of its own, for reasons that are not completely understood MSG brings out flavors that exist in foods, especially the flavor of salt.

Yu Jiamin at Bejing's Sichuan Restaurant said, "MSG is a different flavor than salt but also brings out flavor the way salt does."

As more Westerners visit their country, many Chinese cooks are growing frustrated by what they see as a Western prejudice against MSG. Liu Tong, a cooking instructor of the Sichuan Cooking School in Chengdu, said, "It is not a chemical. It is made from fermentation of cereal. We have always used it in Chinese food."

Actually, the Chinese have not always used it, but the Japanese have. In food history, MSG swam upstream, from Japan to China, instead of the reverse direction of most Asian food. Traditionally, the Japanese got it naturally from a seaweed known in Japanese as *kombu* and in the West as *laminaria*. MSG was first isolated as a substance—a sodium salt of glutamic acid—in a Japanese laboratory in 1908. Since the 1950s, it has been made by fermenting wheat gluten.

Liu Tong said that MSG was needed because Chinese food does not directly use salt.

⌒

THERE ARE NUMEROUS Chinese salt and bean condiments such as douban and dousi, and the Japanese have their own assortment. But the most important is the ancient soy sauce. In China, schoolchildren learn a jingle from the Middle Ages with the seven necessities needed every day: firewood, rice, oil, salt, soy sauce, vinegar, and tea.

In China, there is an ancient tradition of soy sauce made by peasants, but such sauce is becoming a rarity. Today, in both China and Japan, soy sauce is made in factories. Most Chinese say that it is a complicated process and the factories do it as well as the peasants ever did. Anyone who has tasted the thick peasant product might dispute this. Huang Wengen, for one, said the old farm product was incomparably better. The farmers in Dayin said they stopped making soy sauce in the early 1990s, even when they were still pumping brine with foot pedals. They said it was too much work and that factories sell it so cheaply, they could not compete.

But by a strange twist of economics, an artisanal soy sauce is still made in the Sichuan town of Lezhi. Lezhi is a provincial town whose main street has almost no traffic other than busy little three-wheeled bicycle rickshaws. And yet most of the old buildings have been torn down and replaced with what is becoming China's ubiquitous white tile architecture. At night it looks as if a tricycle gang has taken over the deserted streets of an abandoned housing project.

The Lezhi Fermented Product Corporation was a private factory that was nationalized after the 1949 Communist takeover, known here as "the liberation." The state factory made an industrial soy sauce. But in 1999, in a fit of privatization, the state announced that it was no longer going to produce soy sauce in Lezhi. Since no one was interested in buying the company, its 100 workers were given severance pay and left jobless. Ten of them used their settlement money to buy the company. In order to get operating capital, they sold the large downtown plant and moved up a three-flight outdoor mud-and-stone stairway to a storage area on a hilltop at the edge of town.

They no longer had the equipment or the capital to be an industry. So they decided they would have to make their soy sauce the way peasants used to make it. Xu Qidi, the general manager, said, "We had to start all over. This is the old way to do it."

Factories use the crushed refuse from soy oil production for making soy sauce, but the new Lezhi company uses fresh whole beans that are steamed until soft. The beans are then placed in a storage room on flat, round, straw trays that are about four feet in diameter. Yeast is then added. The trays are left on bamboo racks in the concrete storage room for three days, until mold forms on top.

At this point, factories speed up the fermentation process by delaying the addition of salt and keeping the beans in heated bins. But in Lezhi the moldy beans are mixed with water and salt and stored in big, three-foot-deep crocks. The pots are left outdoors to ferment for six months to a year or longer, depending on weather conditions. When it rains, they are covered with cone-

 shaped lids made of sewn palm fronds. Eventually, the paste looks like mud. Water is added, and the mush is slowly filtered through piping. Then it is sterilized by steaming.

Some sauces are darker, some lighter, some thicker, some thinner. The best Lezhi sauce is not quite as thick as its number two product, but it is black, caramelly, and complex. Differences in sauces are determined by the length of fermentation and the amount of water added at the end.

In Lezhi, soy sauce is still sold the old-fashioned way: Customers bring their own bottles, and the sauce is ladled out of crocks. But it is also marketed under the label Wo Bo, which is the name of a local bridge. A room in the dank little factory has a shiny new machine, the only shiny new thing in the plant. The handful of people that is the company, some in suit and tie, others in workers' clothes, all entrepreneurs of the new China, look on with excitement as this machine seals soy sauce into plastic bags to be sold out of town.

CHINA IS CHANGING quickly. The gray and red courtyard buildings of Beijing, some 500 years old, are being torn down at a pitiless rate. In the glare of neon lights that now explode in the sky every night on top of the capital's new high-rise buildings, are Kentucky Fried Chicken and McDonald's advertisements. Fried chicken has been easier to sell than hamburgers, because the Chinese have been eating fried chicken for centuries.

Guo Zhenzhong, a sixty-three-year-old professor who lives in a small apartment stacked floor-to-ceiling with books in a ten-year-old apartment block that already looks eighty years old, does not fit perfectly into this new China. He dresses simply, studies his books, takes great pleasure in traveling to international academic conferences, and appears not to have heard the news that China has switched to a market economy. He cares little for the new consumerist China with its Western labels, both real and fraudulent. Like most Chinese, he still eats the old foods. He once went to a McDonald's. What did he think of it?

He shook his head disapprovingly and said, "No vegetables."

~

CHAPTER TWENTY-FIVE

More Salt than Fish

THE IDEA THAT salt enhances the taste of sugar has not entirely vanished from the West. It is a guiding concept of the snack food industry. A clear example of this is honey-roasted peanuts, but in fact salt and sugar are ingredients in most industrial snack food.

Before refrigeration, when butter was preserved with considerable quantities of salt, sugar was thought to counteract and even mask the saltiness.

Since tasting is all that is needed to detect oversalting, some merchants try to mask this taste by adding a little sugar. So in tasting salted butter, if you detect a sweet or sugary taste, don't buy it.—*Francis Marre:* Défendez votre estomac contre les fraudes alimentaires *Protect your stomach against food fraud, Paris, 1911.*

Curiously, the concept of sugar counteracting salt still flourishes in Sweden, a country which imports both its salt and its sugar, and perhaps for that reason gives them equal regard. The first record of sugar in Sweden is from 1324, when, for a funeral of the wealthiest man in the country, 1.5 kilo sugar, 1.5 kilo pepper, .5 kilo saffron—all exotic luxuries—were imported.

According to Carl Jan Granqvist, a well-known Swedish restaurateur and food commentator, "Sugar brings out the saltiness of salt." Cakes are made with salt. Breads are made with sugar. In September, when crayfish are in season in Sweden, they are served with salt, sugar, and dill. Sugar and salt is a leitmotif of Swedish cooking. There is even a Swedish word for it, *sockersaltad*, sugar salting, which is also the first ingredient listed on many labels.

For newcomers to Scandinavia, one of the more infamous uses of sockersaltad is *salt lakrits*, salted licorice candy, which sometimes comes in the shape of herring, sometimes in laces, or in a gumdrop shape, called a salt bomber, with salt sprinkled on top. A salt lakrits–coated vanilla ice cream, sold on a stick, is a *lakrits puck,* though the manufacturer, GB Glace, said it was made with ammonium chloride, not sodium chloride, which does not seem at all reassuring. Swedes often mention salt lakrits as the one thing they miss when they go abroad. Other Scandinavians and the Dutch are afflicted with the same craving.

Also high on the list of foods missed by Swedes abroad is *kaviar*, a name which purists would see as a travesty, since it contains no sturgeon eggs. Kaviar is salted cod roe mashed with potatoes and sold in a squeezable metal tube. The first ingredient listed on the tube is sockersaltad.

The leading use of sockersaltad, and probably the one that has kept the taste in the northern palate, is for curing fish. On the west coast of Sweden, herring is ground with onions and made into fritters, which are served with a sweet currant sauce. One of the most celebrated expressions of this Swedish taste is gravlax, literally buried salmon. Originally, gravlax was salmon that was cured by being buried in the ground for days or months, an old Scandinavian technique used for preserving herring as well. The longer it is buried, the longer it will keep. But, paradoxically, the longer it has been buried, the more it resembles in smell and texture something rotten. Older Icelanders still horrify youth with smelly little chunks of *hákarl*, buried Greenland shark. Burying produces a very smelly fish rejected by most of the public. The Swedes have maintained the popularity of gravlax by replacing it with salmon cured with salt and sugar.

HERRING STILL COMES and goes in the Baltic and the North Sea in ways no one can predict. The sea between Norway, Denmark, and Sweden is called the Skagerrak. Klädesholmen, a flat rocky island only yards off the Swedish coastline in the Skagerrak, has had only six good herring runs recorded in all of history. The first was in the sixteenth century. Then the herring went away and did not return until 1780. From 1780 to 1808, Klädesholmen was awash with herring. The villagers boiled herring in water, and the oil that rose to the surface lit the streetlamps of Paris and London.

In those years, while herring seemed to be vanishing from the Norwegian coastline, the large population on the tiny island of Klädesholmen were fishing and processing herring, as well as cod and ling. The two boiling plants, owned by wealthy Göteborg and Stockholm merchants, made oil twenty-four hours a day. Then, in the early nineteenth century, fewer herring showed up each September. Some blamed this on the foul smell of the island with its herring oil plants dumping stinking waste back into the sea. Klädesholmen smelled so bad, it was said, that even the herring couldn't stand it. There was not another good run until 1880 to 1900, and there were none in the twentieth century.

More than 1,000 people lived on the island in the eighteenth century. By the twenty-first century, the island had only 470 inhabitants. In the early twentieth century, hundreds of women had been employed cutting up herring for canning, wearing aprons made from Cuban sugar sacks waterproofed with linseed oil. Men mixed salt for the herring with Cuban sugar, then added sandalwood, ginger, cloves, mace, coriander, cinnamon, allspice, oregano, dill, and bay leaf.

In the 1980s, there were 200 herring workers. Today, more her-

Women preparing herring for canning in Klädesholmen in the 1920s.
Klädesholmens Museum

ring is produced with fewer than 100 people working for eight herring canneries, all of them family-owned businesses. For the small companies on Klädesholmen, it is more economic to buy barrels of fish from one of two companies that now process herring for all of Sweden. They buy herring and cure it in a brine in which, for every thirteen kilos of salt, nine kilos of sugar are added.

At the big herring plants, decisions about these formulas are still made by the brine mixer, who, like the master salters on the old-time cod vessels, has the highest salary.

But life has changed in all of Sweden. Until a lumber boom in the mid–nineteenth century it was one of Europe's poorest countries. Before the 1960s and 1970s, the only refrigeration in a Swedish kitchen was cabinets with holes to the outdoors in the wall. Historically, salted provisions got Sweden through its long winters, and traditional Scandinavian food is very salty. A Swedish sausage is coated in a white layer of salt. In Sweden, Äppelfläsk used to be made in the fall when the apples came in. They were sliced and sauteed in salt pork and sugar syrup. But today, everywhere in Scandinavia, as in much of the rest of northern Europe and North America, people are eating less salt and less salted foods. Few eat Äppelfläsk anymore.

Some of the disappearing uses of salt seem so strange it is difficult to understand why they were ever popular. *Snus* is tobacco and salt. It was molded into a wad with the fingers, jammed up between the cheek and gum, and sucked on for an entire day; fresheners were added every hour or so. Some even put it in before going to bed and woke up at night for a fresh wad. And even the Swedes wonder at the habit of the Laplander in the far north, drinking salted coffee.

SALT CONSUMPTION IS declining in most of the world. The average twentieth-century European consumed half as much salt as the average nineteenth-century European. But there is still a love of salt cod, herring, hams, sausages, olives, pickles, duck, and goose preserved in salt—foods that are no longer necessary. Salt cod is sometimes sold only slightly salted so that it requires less soaking, though this convenience is at the expense of quality. Some salt cod is so lightly salted that it is kept frozen, which makes little sense economically or gastronomically. Bacon and salted beef remain popular but, because they are now refrigerated, are no longer so salted that they need to be soaked before using. Since salt curing has lost its function as a way to preserve meat, the paradoxical notion of "fresh ham" has appeared. By Swedish law, a ham that is salted in September cannot be called a "fresh Christmas ham." But a ham that is frozen in September and thawed and salted on December 17 is a "fresh Christmas ham."

In North America, the Jewish delicatessen is a citadel of salt-preserved foods—foods that could just as easily be purchased fresh, including pickled cucumbers and tomatoes, salted and smoked salmon, carp, whitefish, and sable, and cured meats such as tongue, pastrami, and corned beef. Pastrami, of Romanian origin, is dried, spiced, and salted beef, smoked over hardwood sawdust and then steamed. The name may come from *pastra*, the Romanian verb "to preserve." It is available in every delicatessen, but most famously as the specialty of Schwartz's in Montreal. Schwartz's and its pastrami is such an institution in Montreal that after the controversial 1977 Bill 101 required store names to be in French—a language that does not use apostrophes—Schwartz's was one of the few allowed to keep its apostrophe. But it had to change from being a "Hebrew delicatessen" to a *charcuterie Hébraïque*.

While the Jews and the delicatessens are concentrated in eastern North America, much of their fish is taken from the Pacific. Great Lakes carp is becoming rarer, and an inexpensive substitute was salted and smoked sable or sablefish, a huge, deepwater Pacific fish. In the Pacific Northwest, it is known as black cod, though it is nothing like a cod and belongs to a uniquely northern Pacific family. Now that it has become fashionable in the United States and Japan to eat black cod, it is becoming rarer too and cured sable is no longer inexpensive.

Though curing salmon is an ancient tradition everywhere that the fish is found, Jews who learned of it in Germany and central Europe, where it had long been a popular food, did much to popularize it in the world. It was through the Jewish immigrant neighborhoods of Paris after World War II that cured salmon became a staple item of Paris charcuteries. In New York also, it was the early-twentieth-century central European Jews of Manhattan's crowded Lower East Side who first established cured salmon as a New York food and then an American food.

The popular Jewish cured salmon was called *lox*, Yiddish for salmon, from the German *lachs*. Lox is salt-cured salmon, usually Pacific salmon. In the nineteenth century, the Pacific Northwest became a leading center for cured salmon for both the East and West. The booming fur trade of the region bought large quantities of salt. The merchants in the Northwest found that salted salmon sold well in the world and that the ships bringing in salt could buy salted salmon for their return cargo.

Hawaii was a salt supplier for the Northwest. Like many Pacific islands in the eighteenth and nineteenth centuries, Hawaii had an important trade provisioning whalers and other ships with salt and salted meats. Hawaii produced sea salt in inland lakes, the most famous of which was a volcanic crater rumored to be bottomless—evidently not true since the drained salt lake is now in Honolulu filled with high-rise buildings. Hawaiians traded salt in the Northwest and in turn bought salted salmon, a hard product that required soaking like salt cod. Hawaiians still mix soaked salt salmon with tomatoes, a dish they call *lomilomi*. The word means "massage" and refers to the process of flaking the salt fish.

Salt-cured lox, once the leading cured salmon, has in recent years been almost completely abandoned for the less salty Nova, a lighter cure, soaked in brine and then smoked. In recent decades there has been a mantra among Jewish shoppers, "Get the Nova; the lox is too salty." The name comes from Nova Scotia, though most Nova originally came from the nearby Gaspé Peninsula of Quebec. Now there is western Nova, made from Pacific salmon, because Atlantic salmon has all but disappeared except for farmed varieties. Moe Greengrass, owner of a popular Jewish smoked fish store on Manhattan's Upper West Side started by his father in 1929, said, "Nobody buys lox anymore—we sell 100 pounds of Nova and 5 pounds of lox per week." Moe's father, Barney, who had worked in fish stores on the Lower East Side before opening his West Side store, was one of those who had made lox a New York food back when New Yorkers liked their fish salty.

⌒

ANCHOVY IS A fish that has remained more popular salted than fresh, but because salting is no longer a necessity, it has become considerably less salty. J.-B. Reboul, the nineteenth-century Provençal chef, is credited as one of the first to use anchovies creatively, inventing anchovy patés and several pastries with anchovy fillings. He also wrote one of the great recipes of a Provençal classic: anchoïade.

After having washed seven or eight anchovies, let them soak several minutes in water to desalinate; having separated the fillets from their bones place them in a dish with several spoonfuls of olive oil, a pinch of pepper, two or three garlic cloves chopped fine, you could also add a splash of vinegar.

Cut a slice about one inch off the top of a *pain de ménage* [or *pain ordinaire*—a long, round, typical French bread]. This is the best choice of bread because it does not easily crumble.

Divide this long slice of bread into two or three pieces: they should be the same. Make one for each guest. Place some anchovy fillets on each piece and arrange the pieces in a dish.

Cut the remaining bread into small squares. Everybody dips the squares in the prepared oil and then uses the square to crush the fillets on the bread. When it is all crushed together, anchovy and sauce, you eat the squares of bread that were used for crushing while toasting the slices with crushed anchovy on top; it releases typical flavor that fills with joy all lovers of Provençal cooking and gives pleasure to many a gourmet.—*J.-B. Reboul,* La cuisinière Provençale, *1910*

Another celebrated nineteenth-century Provençal chef, M. Morard, wrote, "The laziest of stomachs and the sleepiest of appetites are obviously forced to awaken at the first mouthful of this stimulating slice of bread, made golden with olive oil, awaiting crushed anchovy fillets and chopped garlic, that the culinary mosaic-maker has so perfectly placed on top."

⌒

IN 1905, HENRI Matisse and André Derain went to Collioure, the little pink-and-yellow village by the sea, still famous for its anchovies. In one of the most fruitful summers in the history of art,

they produced paintings of furious colors. Derain painted a village of pure primary color and Matisse a village of vibrant opposites, turquoise and orange, magenta and gold. They took their paintings to Paris's Salon d'Automne that year and created a sensation, a movement in the art world known as fauvism.

Visit the little port of Collioure today, a few miles up the Catalan coast from the Spanish border, and these works of Matisse and Derain will seem purely imaginary. Collioure is no longer a world of brilliant colors but subtle pastel walls where wisteria blooms pale purple and magnolia pink.

What is missing are the fishing boats.

Derain painted them red and yellow, their bright red masts poking out of the harbor like an autumn grove. Matisse depicted them in a red bunch seen from his turquoise window. They were exaggerating the colors, but the fishing boats, called catalans, truly were painted in blazing primary colors. These were the boats of anchovy fishermen.

In 1770, Collioure had 800 fishermen working on 140 catalans. In 1888, the number of boats had declined by ten. The fishermen observed that anchovies will rise near the surface on a night with a full moon. Reasoning that the fish were attracted by the moon, they started making their own moons once they had electricity, and they called them *lamparos*. A lamparo was a huge light on a buoy that was some five feet high, carried out to sea hanging from a hook mounted at the bow of the boat. On a calm, moonless night, the fishermen would set their nets around the buoys and then turn on a lamparo and wait for the anchovies to gather under it. Then they would haul up a full net.

The catalans would go out every night and bring the catch in every morning for salting. The catches were good, but the more they caught, the farther out to sea fishermen had to go to find anchovies. Fishermen started using big steel-hulled ships for the longer voyage, but such vessels could not dock in Collioure because the harbor was not deep enough. By 1945, there were only twenty-six working catalans left in Collioure. Today, only one catalan sits in Collioure harbor by the medieval walls. It is an unused souvenir of the village's anchovy industry.

Art lovers, wanting to see the town Matisse and Derain painted, flock to Collioure for the tourism season, which, like the old anchovy season, is May to September. But the colors, the fishing boats, are not there. The locals still make Banyuls from their vineyards in the winter, and in the summer, instead of fishing, there is the tourism. Two families still salt anchovies in Collioure, using salt from Aigues-Mortes. The anchovies are caught in Port Vendres, a contemporary hillside monument to industrial efficiency. It has a fleet of vessels that find the schools with sonar and an afternoon fish market that auctions the catch. But Matisse would not have painted it.

A NUMBER OF seventeenth-and eighteenth-century writers insist that the heir to garum is not anchovy sauces that later became ketchup, but salted fish eggs, caviar. Guido Panciroli, who wrote a 1715 book titled *History of Many Memorable Things Lost*, believed, "Other types of garum called bottarga and caviar now take their place."

The word *caviar* is of Turkish derivation and refers to the eggs of the sturgeon—a prehistoric animal that has not evolved in 180 million years. It is a huge migratory fish that, like the salmon, is anadromous, that is, it lives in saltwater but swims upstream to spawn in the freshwater place of its birth. The eating of its eggs may well be as old as or even older than the eating of garum. Originally, the eggs were food for fishermen, cheap food because they were not salable, whereas the fish itself brought high prices. But gradually the eggs gained appreciation. In 1549, their preparation was described by Cristoforo di Messisbugo, a Renaissance food writer from Ferrara:

CAVIAR TO EAT FRESH AND TO CONSERVE
Take sturgeon eggs, the best of which are black. Spread them out on a table using the blade of a knife. Take out the ones that are filmy, weigh the remaining ones and, for every 25 pounds of eggs, add 12 and one half ounces of salt, or one half ounce per pound.

The medieval rivers of Europe were full of egg-bearing sturgeon. They were common in the Seine, the Gironde, the Thames, the Po, the Danube, the Ebro in northern Spain, and the Guadalquivir in southern Spain. The fish were often a subject of royal privileges. The British monarch, starting with Edward II in the thirteenth century, claims the right to the first sturgeon caught in British waters every year.

France produced caviar since at least the time of Louis XIV, largely from the sturgeon catch on the Gironde River. But sturgeon caught on the Seine, even in Paris, was a rare enough event for the fish to be presented to kings. Colbert regulated the fishery to preserve the fish, and these laws are still in force. But the fish are gone. Louis XV got a Paris sturgeon in 1758 and Louis XVI got one in 1782. Antonin Carême, the famous early-nineteenth-century French cook, insisted he saw a 220-pound sturgeon almost three yards long by the Pont de Neuilly on the western edge of Paris. That was one of the last sturgeon sightings in Paris.

Sturgeons, which can weigh up to two tons, have little resistance to industrial pollution. Even the Gironde, the last holdout of French sturgeon, became too polluted, as did the Hudson and other great sturgeon rivers of North America.

When Europeans settled in North America, they recorded seeing Native Americans catching huge sturgeon. Even in the nineteenth century, American rivers had sturgeon. Caviar was served as a free bar snack, in the hope that as with peanuts, the saltiness would encourage drinking. During World War I, British soldiers were fed cans of pressed caviar, which they called "fish jam" and mostly loathed. A soldier would pay for cans of sardines rather than eat the free fish jam that was issued.

For caviar to have been considered the heir to garum, it had to have been used as a seasoning rather than being eaten by itself. Until the twentieth century, that seems to have been the case. In nineteenth-century Russia, sauerkraut was valued more than caviar, and in this recipe the caviar is simply a pleasant salted flavoring for the cabbage:

SOUS IZ KAPUSTY S IKROJ
(SAUERKRAUT WITH CAVIAR)

Boil three pounds shredded sauerkraut, adding only enough water to prevent it from burning. When it has cooked, drain in a coarse sieve. Melt a half pound Finnish butter in a skillet, stir in the sauerkraut, and fry both together. Pour salted water over caviar that has just been removed from a fresh fish, mash it fine, mix with the sauerkraut, and let it boil thoroughly over the fire. The caviar will impart a fine flavor to the sauerkraut. This dish may be served with patties, small sausages, or fried fish. On fast days, substitute olive oil for the Finnish butter. Add enough caviar so that the sauerkraut appears as if it were strewn with poppy seeds.—*Elena Molokhovets,* A Gift to Young Housewives, *1897*

Apparently, by the early twentieth century, Americans valued Russian caviar from the Caspian Sea. In 1905, Russia was in open revolt against the czar, and in April 1906 an American publication, *Wide World Magazine,* warned, "The unrest in Russia, it is feared, will greatly affect the caviar industry." The writer was concerned that the cossacks would get involved in the political unrest and abandon fishing sturgeon.

But the magazine did see reasons for revolt. It reported that the czar forced the cossacks to give him a yearly tribute of eleven tons of their best caviar and reported that this tribute alone required the killing of 5,000 sturgeon at the start of each season.

The article described the cossacks in the Russian winter, standing on the ice and fishing through holes. It reported that the eggs were mixed with "the finest salt," at a ratio of 4 to 5 percent salt. The Caspian, fed by the Ural and Volga Rivers, is the world's largest saltwater lake. Not only does it have sturgeons, it has sea salt where brackish water evaporates at the mouths and estuaries of the numerous rivers that empty into it.

Before the 1917 revolution, the cossacks were the dominant caviar producers of the Caspian. They fished sturgeon only twice a year, for two weeks each time, and the sturgeon seemed inexhaustible. The entire cossack population participated in these two brief fishing seasons. First, in the autumn, whole extended families pulled nets down the Volga River. The second two-week season was in the middle of winter, and this fishery also was on the river. Armed with harpoons, hundreds of cossacks would stand on the ice of the frozen river awaiting a cannon blast that was their signal to pierce the ice and attempt to spear a sturgeon. The noise would drive the terrified fish downstream, and the cossacks would follow with their harpoons and cannons while merchants from Moscow, Leningrad, even Paris and other European capitals waited for the giant fish to be cut open while still alive.

The price has been leaping upward since the beginning of the twentieth century. From 1900 to 1915, the price of caviar doubled. Merchants began importing Russian caviar, not to mention French caviar with Russian labels, to exclusive establishments in western Europe. This was how the Petrossian family, today one of the leading caviar distributors, got started. Born on the Iranian side of the Caspian, they grew up on the Russian side, immigrated to Paris, and discovered that Russian things were in vogue with the rich.

During the twentieth century, as industrial pollution and oil spills killed off sturgeon around the world, commercial caviar fishing was largely reduced to the Caspian Sea. Historically the Caspian has always been controlled by Russia from the northern shore and Iran from the southern shore, giving these two nations a virtual monopoly on caviar. But the Caspian and the Russian rivers that feed it have also been besieged by pollution. Chemicals and fishermen have killed so many sturgeon that by the early 1970s even the Russians were suffering a shortage of caviar. In that decade, increased industrialization in Iran started threatening the Iranian fisheries on the southern side of the lake. At this point, the price of caviar became prohibitive to most people.

Sturgeon is still fished in the Caspian only twice a year, with a larger catch in the spring and a smaller catch in the fall, hauled by net from its habitat in the deepest pockets of the Caspian floor. Once landed on a factory ship, the sturgeon receives, as do many fish caught by man, a clunk on the head. The blow knocks the fish unconscious and permits the fisherman to open it live and remove the roe.

The eggs must be passed through a wide screen to separate them from other fibrous matter, and then they must be delicately mixed with salt, both to preserve the eggs and to bring out the taste. As is often the case when preparing food, there is a delicate trade-off: The more salt, the better preserved, but the less salt, the finer the taste. Preparing the caviar takes careful labor, which also increases the cost. In general, Iranians salt their caviar less than Russians, although each fishery is slightly different, and in fact each fish is different, which is why the eggs in any sized can of caviar are all from the same fish.

Of the twenty-four known varieties of sturgeon, three are still fished for caviar in the Caspian. The prices of the caviar from the three varieties—beluga, ossetra, and sevruga—are not a reflection of quality but rather of the rarity of the fish. The giant beluga are hardest to find, and therefore their caviar is the most expensive. It takes twenty years for a female beluga to mature, and at that point she can weigh as much as 1,800 pounds and be up to twenty-six feet in length. Such a fish could yield twenty pounds of eggs. Beluga have the largest eggs, and these smoky gray bubbles are also the most delicate eggs, which is another reason they are the most expensive. More beluga eggs are broken and lost in processing than any other kind of caviar.

The dominance of Caspian caviar has been declining, and it now represents only about three-quarters of the world's supply. With Caspian caviar growing scarcer and more expensive every year, there have been attempts to bring back European and American caviar, partly through farmed sturgeon, yielding such scrupulously labeled products as the "American Paddlefish caviar."

⌒

BUTARGHE

Take the eggs from a fresh mullet, best when in season, and take care not to break the delicate skin surrounding each egg and add a discreet amount of salt to the eggs—neither too much nor too little—and leave this way for a day and a night. Then place it in smoke far enough from a flame so that it does not feel the heat. When it has dried then place it in a wooden box or barrel surrounded in wheat bran. This bottarga is typically eaten uncooked. But those who want it cooked can heat it under ashes or in a clean, warm oven, turning it, but just until it is hot.—*Martino,* Libro de arte coquinaria *(Book of the art of cooking), 1450*

LITTLE IS KNOWN of Martino, including his full name. He was born in Como, worked for aristocrats, and was one of the most respected and influential Italian cooks of his day. His bottarga recipe calls for mullet eggs and is smoked rather than pressed. Bottarga, which the eighteenth-century Panciroli had listed with caviar as a possible descendant of garum, varies with the fishery. Native Americans used to make it by pressing and drying sturgeon eggs. Today, it is usually salted, pressed, and dried fish eggs. In Tunisia, bottarga is made from mullet eggs and is a product associated with the Jews of Tunis, the same way smoked salmon and pastrami are associated with Jews in North American cites. But the name is thought to come from the Arab *bitârikh,* and it was also made in ancient Egypt, probably also from mullet.

Today in Italy, bottarga has come to be thought of as a Sicilian food, specifically from western Sicily, and that means tuna, not mullet, eggs. The tuna trade on the west coast of Sicily combines one of the oldest saltworks in Europe and one of the oldest tuna fisheries. Between the two, the port city of Trapani juts out on the triangular tip of a narrow peninsula. Typical of Sicilian towns, Trapani has a Phoenician-Roman-Norman-Arab-Crusader history. These elements are reflected in the architecture, the language, the food, and the customs. Everything in Sicily is built on layer after layer of history—all the people who came, conquered, built, were defeated, and left.

The Castiglione tuna company, just north of Trapani, makes more than 2,000 pounds of bottarga every year, which Sicilians grate over spaghetti with olive oil, garlic, and chopped parsley. The eggs come from the bluefin tuna that enter the Strait of Gibraltar once a year and swim past western Sicily to their Mediterranean spawning grounds. Each female has two huge roes weighing between six and seven pounds each. Workers prepare a brine from the local sea salt and wash the roes in it and then cover them in the coarse-grained salt that is a regional specialty. They then place a thirty-kilogram (sixty-six pound) weight on the salted roes. More weight is added every week until, by the end of a month, sixty to seventy kilos, the weight of a middle-sized man, are pressing on the salted roes. After pressing, the roes are dried in the sun for a week.

Like the sturgeon fishermen of old, the Sicilians sell off the fish and eat the eggs. But they also sell the bottarga all over Sicily. They sell the tuna hearts as far away as Palermo, the Sicilian capital. *Lattume*, the delicate-tasting salted male reproductive gland, is for locals in the Trapani region, as are tuna intestines, stomach, and esophagus.

For centuries, this coast was famous for its salted tuna as well. But these days Sicilians don't eat their bluefin tuna in any form; they sell it fresh for dazzlingly high prices. Ninety percent of the local catch is landed one hour after being killed and instantly sold and flown to Japan.

The passage of the bluefin off the Mediterranean coast at spawning time was first observed by the Phoenicians, who set up what is called in Sicilian *tonnaras*. As various cultures became dominant in this passage between Sicily and Tunisia, the tonnara became layered in ritual. Today it only continues in two places in Sicily, the little waterfront town of Bonagia just north of Trapani and the small nearby island of Favignana. The tonnara in Bonagia is owned by the Castiglione company, which usually, despite the high prices from Japan, loses money on it. The bulk of its profit comes from yellowfin tuna caught elsewhere and bought and packaged by the company. The bluefin is vanishing not because of the tonnara but due to far more efficient fisheries in the Atlantic. Eugenio Giacomazzi, the Castiglione production manager, said that 1,000 pounds of bluefin is now considered a good catch. That used to be three or four fish, one fish according to ancient accounts, but today it is a netful, because fish species, as they become scarce, mature younger and become smaller animals.

Castiglione has 150 employees. But every March another 120 are hired to work the tonnara. The leader is known by the Arab word *Raiz*, and the fishermen sing an Arab song, "Cialome" (pronounced SHALOMAY), to invoke the gods for the hunt. But the final kill goes by the Spanish word *matanza*, which, appropriately, means "slaughter."

The tuna hunt begins in March with men on the narrow waterfront of Bonagia repairing and arranging nets as they sing traditional songs, part in Sicilian and part in Arab. Instead of exhausting the fish on the end of a line, in the Sicilian tonnara the bluefin is worn out by being led through a series of nets over a number of days. A net wall 150 feet high, four and a half miles long, is anchored to the ocean floor running east to west. In May and June, the tuna enter the Mediterranean. Approaching the coast of Sicily, they turn south to pass through the straits between Sicily and Tunisia but instead hit the net wall and run along it into what is called "the island," which is a series of net rooms. In ancient times, the large fish were guided through the rooms by men with long sticks. Today, this is done by a scuba diver known as the big bastard. The bastard of Bonagia, Maurici Guiseppe, said that as he swims with the ill-fated fish, he passes amphorae and other ancient artifacts of shipwrecks from the Phoenicians, the Greeks, and the Romans.

The big bastard's job is to coax the tuna from one room to an-other—each of the rooms has a name—until, after about two weeks, the fish are exhausted, awaiting their fate in the *camera di matanza*, the slaughter room. The net is hauled up, and fifty-five fishermen in a long boat spear and gaff fish. It is an ancient way of fishing tuna. Twenty-five hundred years ago, in *The Persians*, Aeschylus, describing the Greek destruction of the Persian Navy, said it was like slaughtering tuna. The large bluefin, even though tired out from the weeks of manipulation, thrash and struggle. The Mediterranean turns black with their blood, and the foam of the water turns scarlet as they are stabbed, gaffed, landed, and shipped to Japan.

The tonnara fishermen spend March repairing the nets, April  setting them up on the ocean floor, May or June fishing, then taking the nets back up. Then it is July, time to work the salt harvest.

SOUTH OF TRAPANI along the coast, earthen dikes begin to appear, and a few stone windmills. The dikes mark off ponds, some of which hold turquoise water, some pink. The stone towers of windmills stick out from these orderly pastel ponds. The saltworks are built out along the coast until, toward the south, deep green leafy fields take over, which are the vineyards of Marsala wine. This is one of the oldest salt-making sites in the world—the one started by the Phoenicians to cure their tuna catch, and after the destruction of Carthage, continued by the Romans. When the Muslims were in Sicily from 800 to 1000, they wrote of the windmills of Trapani.

The current windmills are based on a Turkish model that was adopted by the Spanish, who brought their windmills to Sicily and later to Holland. About the year 1500, windmills were built here by a man named Grignani to move brine through the ponds. His son was named Ettore, which is the name of these saltworks facing the isle of Mozia. Until the saltworks on Mozia were destroyed by the Romans in 397 B.C., the Carthaginians had made salt there as well.

Trapani salt was sent to the Hanseatic League in Bergen and was known throughout medieval Europe. But the unification of Italy in the nineteenth century made it difficult for Sicily to market its salt. The Italian government had a salt monopoly, and it protected its saltworks in Apulia by not allowing Sicilian salt onto the mainland. The monopoly was resented because it made salt expensive. In his 1891 book, *The Art of Eating Well*, Pellegrino Artusi, the Florentine silk merchant turned popular food writer, suggested for his ice cream recipes:

> To save money, the salt can be recovered from the ice water used to freeze the ice cream, by evaporating the water over a fire.

For centuries an important export, Trapani salt became a local product, used to cure the tuna catch, lavishly sprinkled on grilled fish in the Trapani area, and to preserve the caper harvest. Capers are the buds of *Capparis spinosa,* so spiny, in fact, that the Turks call it cat's claw. They grow wild in the Roman forum but are so tough that they also grow between cracks in the rocks in Israel's Judean desert. They seem to love rocks and grow along the coastline boulders of southern Italy and Sicily, brightening them with purple and white flowers. But the buds must be picked before they begin to open, which requires a daily harvest in the summer and a careful examination of each bud. Then the buds must be cured to bring out the characteristic flavor. The French consider their best capers, which come from Provence, to be the smallest, and they pickle them in vinegar.

In past centuries, when Mediterranean products were difficult to get in northern Europe, nasturtium buds were used as a substitute, as in this English recipe.

Nasturtium Indicum. Gather the buds before they open to flower: lay them in the shade three or four hours, and putting them into an earthen glazed vessel, pour good vinegar on them, and cover it with a board. Thus letting it stand for eight or ten days: then being taken out, and gently press'd, cast them into fresh vinegar, and let them so remain as long as before. Repeat this third time, and barrel them up with vinegar and a little salt.—*John Evelyn,* Acetaria: A Discourse of Sallets, *1699*

In Sicily, the prized capers are large and come from the island of Pantelleria, which is part of Sicily, though closer to Tunisia, or the tiny Sicilian islands off the north coast. Sicilian capers are kept in coarse Trapani salt, often without vinegar, soaked before using, and served with grilled fish.

Confined to their island market by the Italian monopoly, Sicilian traditions—tuna, capers, table salt, the island's ample olive production, cheeses, and sausages—were not enough to sustain the saltworks, and most were abandoned by the 1970s.

But not the saltworks of Antonio d'Ali, who had kept his alive at a third of capacity for years. "We kept going because we always hoped that the monopoly would be dropped someday," he said. In 1973, the Italian government ended the salt monopoly, and today these saltworks provide salt throughout Italy, which was what the Italian government had feared in the first place.

⌒

COMPETING WITH TRAPANI for longevity are the saltworks the Phoenicians started at Sfax, across the Sicilian straits in Tunisia. Today, Sfax is an unglamorous industrial town with three- and four-story buildings and a few palm-lined boulevards. It is Tunisia's second largest port after Tunis but the leading port for phosphates, olive oil, and salt.

The saltworks are located along the coast south of the port. In late winter, when the salt crystals are being scraped up and harvested, the area is fragrant with the white blossoms of nearby almond groves. Small vegetable gardens are fenced off with rows of prickly pear cactus, which the Spanish brought from America and are now part of the North African landscape. As with most everything in Sfax, a visible layer of brown dust dulls the skins of the cactus and the vegetables. It is the encroaching sand of the Sahara. The dusty vegetables include cauliflower, carrots, and cucumbers, which are put up in salt brine.

Built out into the sea are 3,000 acres of evaporation ponds and another 400 acres reserved for the final crystalization, which produces 300,000 metric tons of salt per year. Noureddine Guermazi, the plant director, was asked if salt was a major part of the Tunisian economy. He answered with an ironic smile, "In tonnage, yes." Most surviving saltworks in the modern world make a profit on producing huge quantities of salt and transporting it relatively inexpensively.

Sfax is a good salt location because it has only eight inches of rain per year, which makes it much drier than Tunis and the north coast. Europe and eastern North America have more than forty inches of rain in a dry year. To the south is the Sahara, where there are still sebkhas. Sometimes the dry salt beds are harvested with bulldozers. Farther south into Africa, there are places where camels are still used. Taoudenni, in northern Mali near both the Algerian and Mauretanian border, was first described to Europeans in 1828 by René Caillié on his geographic study of the Sahara. He found Taghaza, the city of salt, already abandoned. But in Taoudenni, he reported that rock salt, relatively pure sodium chloride, was found a few feet below the surface. Today the same mine, using the same techniques, is controlled by Moors, tall people clad in sky-blue robes, part Arab and part Berber seminomads from Mauritania. The Moors pay Malians about two dollars a month to dig thick blocks out of the salt crust and pack them onto camels that travel south in caravans of thirty or forty camels to Timbuktu, still a trading center on the Niger River.

Farther east, due south of Tunisia, the gray sands of Bilma, Niger, are pockmarked by random deep pits. The pits are deep from centuries of digging, but more salt is always there. A single family continues digging a hole for generations. Today, the salt is sold for about fifty cents for a thirty-pound block to traders who carry the blocks by caravan—with as many as 100 loaded camels—for more than two months across Niger to northern Nigeria, where Bilma salt is valued for livestock. There the fifty-cent blocks sell for about three dollars. Were the labor not so cheap, no profit at all could be made from such Saharan salt.

But the sea salt at Sfax is loaded onto ships and sold all over the world. A great deal of it goes to that still salt-hungry corner of Europe, Scandinavia, for salting fish and for deicing roads. The fact that salt lowers the freezing temperature of water has given salt producers a huge winter market on northern highways, and this has become a far more important use of salt than fisheries. The salt fish trade has undergone a historic reversal. With the once precious salt crystals so common they are dumped onto roads, today there is a scarcity of tuna, anchovies, herring, Great Lakes carp, Caspian caviar, even cod.

In Sfax for Aïd Essaghir, a Muslim holiday after the fast of Ramadan ends, salted fish is poached and served with a sauce known as *charmula*. Wealthy people sometimes use salt cod, which is imported from northern Europe and increasingly expensive even though it is often cured with salt shipped from Sfax. But most in Sfax salt their own local fish. The saltworks in Sfax sells noticeably more salt to the local market at Aïd Essaghir than at any other time of year.

Charmula is one of numerous Tunisian examples of salt and sweet being used together. But Tunisians say that all these salt-and-sweet dishes are foreign imports brought from Spain in 1492 by expelled Muslims.

The Affes family, which owns one of the two largest couscous factories in the world—the other is in Marseilles—is from Sfax. Here is Latifa Affes's recipe for charmula:

Salt any large fish. Poach it and serve with the following sauce:
1 kilo red onion, 1 kilo raisins, ½ liter olive oil, salt, black pepper (some use coriander powder but I do not).

Mince onions and cook them slowly in olive oil for about two hours. Soften raisins in water and pass through a sieve to remove seeds. Add to olive oil mixture and cook on low heat for two days. Add salt and pepper.

Acres of rock-reinforced dikes mark off the salt ponds at Sfax, which host leggy birds—white egrets and pink flamingos growing pinker as they feed on the brine shrimp, their color reflecting in the milky saltwater. They graze there in the winter, and then, as though following the salt harvest, they fly to the swampy estuary of the Rhône in southern France and graze in the salt ponds at Aigues-Mortes. The flamingos live better than they did when they had to visit Roman saltworks.

Pluck the flamingo, wash it, truss it, put it in a pot; add water, salt, dill, and a bit of vinegar. When it is half cooked, tie together a bouquet of leeks and coriander and cook. When it is almost cooked add defrutem for color. In a mortar put pepper, cumin, coriander, silphium root [a rare plant from Libya much loved and consequently pushed to extinction by the Romans], mint, and rue; grind, moisten with vinegar, add dates, and pour on cooking broth. Empty into the same pot and thicken with starch. Pour the sauce over them and serve.—*Apicius, first century* A.D.

Today, Sfax and Aigues-Mortes and many other saltworks are protected bird sanctuaries. Also, today's chefs disapprove of sauces thickened with starch. The Romans had felt differently.

They particularly liked eating flamingo tongue, which led Martial, a contemporary of Apicius, to write of the birds:

My pink feathers give me my name,
But my tongue among gourmets gives me my fame.

Aside from the same flamingos, the saltworks at Sfax and Aigues-Mortes have something else in common: Both were bought in the 1990s by the Morton Salt Company.

~

CHAPTER TWENTY-SIX

Big Salt, Little Salt

TRANSPORTATION WAS ALWAYS the key to the salt business, and Morton was a company founded on a transportation idea. In 1880, Joy Morton, a twenty-four-year-old Detroit-born former railroad employee, began working for a small Chicago company, E. I. Wheeler and Company. The company had been started in 1848 by Onondaga salt companies to serve as an agent, selling their salt in the Midwest. Morton, whose father, J. Sterling Morton, would later become Grover Cleveland's secretary of agriculture, came to the Chicago company with a small amount of money to invest and an idea. Staking his entire savings, $10,000, young Morton bought into the company and acquired a fleet of lake boats. During the summer when the Great Lakes were completely open and ice-free, his barges could inexpensively deliver a year's supply of salt to midwestern centers. At a time when salt companies were fighting for the expanding midwestern market,

Morton's company, with its Great Lakes barges, had an important competitive advantage.

By 1910, when the company incorporated, it had purchased saltworks, and the Morton Salt Company was now both a distributor and a producer. One of its early innovations, in 1911, was the addition of magnesium carbonate to table salt, which kept the salt crystals from sticking together; as stated on the box, the salt "never cakes or hardens." Eventually, the chemical was replaced with another nonsticking agent, calcium silicate. This nonsticking quality was to become the basis of Morton's famous marketing campaign. Another innovation: In 1924, on the recommendation of the Michigan Medical Association, Morton produced the first iodized salt.

In those years, when the vacuum evaporator was still a new idea and a fascination surrounded the concept of uniform salt crystals, Morton claimed that every crystal it produced was of the exact same size and shape. "The final product is of such uniform, high quality and grain that inspection under a microscope cannot reveal a difference between Morton salt made in New York and Morton salt made in California," the company asserted. Morton bought saltworks all over the country. Some evaporated seawater, others heated brine, and still others mined rock salt, and yet a single consistent product was made that customers identified simply as "Morton's salt."

The company created a cylindrical package and even patented the little metal pouring spout and hired an advertising firm, N. W. Ayer, to launch the first nationwide advertising campaign ever undertaken for salt. Morton commissioned twelve advertisements to run in consecutive issues of *Good Housekeeping* magazine. But rather than taking the twelve, the company seized on one of the ad

The Economy in Good Seasoning

MAKE things taste right in the first place and nothing will be wasted. Morton's Salt helps you save through right seasoning.

You can follow any recipe with Morton's

because it's *all salt*, and so when the recipe calls for one-half teaspoonful you are safe in using just that amount. There's no guesswork in the cooking because there's no powder in the can.

And Morton's Salt *pours* for the same reason—it's *all salt*—perfect cube crystals, free-running. Note the handy can with adjustable aluminum spout. When open it allows the salt to flow easily and without waste. Closed, it protects the contents.

MORTON SALT COMPANY
89 E. Jackson Blvd., Chicago

One of the "big little things" every woman can afford

When It Rains
IT POURS

MORTON'S
FREE-RUNNING
SALT
IT POURS

MORTON'S SALT

From The Ladies' Home Journal, *October 1919.*

agency's backup ideas, a little girl in the rain holding an umbrella and spilling salt. The original slogan was "Runs freely," but then someone suggested "It never rains but it pours," which was deemed too negative and was replaced with "When it rains it pours." The ad, which first appeared in 1914, not surprisingly does not mention magnesium carbonate but instead claims that the reason it pours so well is "it's all salt—perfect cube crystals. Note the handy can with adjustable aluminum spout." In the 1940s, a poll of 4,000 housewives showed 90 percent recognized the Morton brand.

As it became clear that quantity and transport were the keys to profit in a modern salt industry, small producers began disappearing, and companies such as Morton began buying them out to become bigger. In the nineteenth century, more than a dozen salt companies operated in the southern end of San Francisco Bay. During the twentieth century, these companies were consolidated by the Leslie Salt Company. In 1978, Cargill bought Leslie, and today there are only two companies involved in San Francisco Bay salt, Cargill and Morton. Cargill, a food company that is the largest private company in the United States, is the only salt producer left in San Francisco Bay. Morton buys some of this salt for distribution. Both companies have been buying other salt companies for decades, and they have become the two largest salt-producing companies in the world. It was Morton's 1996 acquisition of Salins du Midi, owners of the Aigues-Mortes saltworks, producers of France's leading brand, La Baleine, that made Morton the world's largest salt company.

⌒

IN 1955, MORTON bought the saltworks on Great Inagua Island in the Bahamas, then a British colony. Most of the few people who arrive on this island bump down in a small plane. The single-runway airport is a lonely, windswept place with two U.S. Coast Guard helicopters that work with the Bahamians looking for small planes from South America carrying white powder that is not salt. A few people cluster around the two-room terminal building because it has a television. A woman cuts hair in the other room. A sign by the runway says, "Inagua, the best kept secret in the Bahamas."

The sea is a blue-green that is almost blinding in its brightness. The streets are paved and empty except for dogs and chickens. The occasional traffic is almost always a pickup truck that says "Morton Salt" on the side. Matthew Town, the capital and only real town, is a grid of about a dozen intersecting streets with little green-trimmed, or yellow, or sometimes hot-pink houses interspersed with overgrown empty lots. Twelve hundred people live in this town and a few hundred on the rest of the island. The general store, well stocked with frozen foods, the leading hotel—a two-story house with a turquoise-green picket fence—the drinking water, the electricity, and most everything else on the island comes from Morton Salt. Morton has 200 employees at the saltworks, but many more in Matthew Town.

Great Inagua is a flat limestone-and-coral island. The soil is impregnated with salt from the sea, which leaches into the inland ponds. In addition, the island is favorable for making sea salt because it is sheltered by the large Caribbean islands of Hispaniola, Cuba, and Puerto Rico, and so rarely is hit full force by a hurricane.

The leaching soil turns the water in an inland lake saltier than seawater. Seawater is pumped into reservoirs in the inland lake, where it remains, sheltered from tides, for nine months, concentrating from solar evaporation, while attracting more salt from the soil. Then it is pumped to a crystalizing pond for twelve more months until it is reduced to a three-inch layer of salt crystals.

In the wetlands surrounding the inland lake live 50,000 to 60,000 flamingos, feeding off the brine shrimp in the 38,000 acres of evaporation ponds. Believing that the shrimp aid evaporation, Morton buys additional brine shrimp eggs from the ponds in San Francisco Bay.

Morton produces no table salt in Great Inagua, only crude road deicing salt, an industrial grade for water softeners, a quality product for the chemical industry, and a fishery salt that is bought by cod fishermen in Iceland. Almost none of Great Inagua's salt stays in the Bahamas. Bahamians buy their water softeners, often made from Great Inagua salt, from Florida. For Morton, Great Inagua has the sea salt production closest to the United States' east coast and produces 1 million tons of salt per year. "You have to drive production," said Geron Turnquest, vice president of operations at the Great Inagua Morton saltworks. "It is volume that makes the profit here."

~

MOST SALTWORKS IN the Caribbean either have been taken over by large international companies or have been abandoned. In the nineteenth century, the Turks Island Company was an important international company. It even owned a saltworks on San Francisco Bay. But in 1927, that saltworks was absorbed into the larger American company that became Leslie, which was bought by Cargill. The salt companies of the Turks were not big enough to compete.

A few miles south of Grand Turk Island, Salt Cay is a two-and-a-half-by-two-mile triangle with a sizable part of its interior occupied by abandoned salt ponds, crumbling rock coral dikes, brine still reddish with brine algae, abandoned metal windmills sticking out over the reddish water like homemade scarecrows. The airport is within walking distance of most of the population. Donkeys and cattle left over from better days wander wild, graze, and breed. The largest native animal in the arid, salty landscape is the three-foot-long giant iguana.

This island offers a chance to see a nineteenth-century Caribbean community. Livestock wander the streets. The tin-roofed whitewashed houses, shutters painted bright colors, their foundations surrounded by a row of conch shells, date mostly from the nineteenth century. There are almost no cars; it is mainly bicycles that use the well-worn sparkly roads that are paved with salt. In recent years, golf carts have been replacing bicycles as the leading mode of transportation. They seem well suited to the salt roads.

Under the rules of colonialism, everything goes to and comes from the mother country. In 1870, the colony of Turks and Caicos was asked to send a crest to England so that a flag for the colony could be designed. A Turks and Caicos designer drew a crest that included Salt Cay saltworks with salt rakers in the foreground and piles of salt. Back in England, it was the era of Arctic exploration, and, not knowing where the Turks and Caicos was, the English designer assumed the little white domes were igloos. And so he drew doors on each one. And this scene of salt piles with doors remained the official crest of the colony for almost 100 years, until replaced in 1968 by a crest featuring a flamingo.

In the nineteenth century, Salt Cay had 900 residents. In 1970, three years after the salt industry died, fewer than 400 remained. At the turn of the twenty-first century, 62 adults and 15 children were the only legal residents. Many of the 62 are retired, and the island would have a labor shortage were it not in the sea path between the island of Hispaniola and Florida. Many illegal Haitians and Dominicans land here, sometimes by mistake, and some find work and stay. If nothing else is available, money can always be made by taking rocks off the dikes and sea walls of the saltworks and sitting on the pile, smashing them one by one with a hammer into gravel for building material.

When the salt business died, the remaining merchants still had stockpiles of salt, often in the bottom floor of their homes. Eight- or nine-man Haitian sailboats arrived at the Turks and Caicos to sell mangoes and other items. Before returning home, they would stop off at the abandoned saltworks of South Caicos, Grand Turk, or Salt Cay, where mountains of salt remained, and scoop out a load.

The last powerful hurricane to hit Salt Cay, in 1945, destroyed all but three of the great houses where wealthy salt-trading families lived with their salt stored in the basement. In the late 1990s, one of the remaining three salt houses collapsed from termite

An early twentieth-century postcard showing salt raking in Grand Turk.
Turks and Caicos National Museum, Grand Turk, B.W.I.

damage, exposing its cache of salt. This unprotected, graying, and rain-eaten iceberg is the last of Salt Cay salt, except for a small crusty pile left slowly melting from humidity in the basement of one of the two great houses still standing.

Native Salt Cay residents are called "belongers." Many of the belongers were salt workers, who left for merchant marine service in World War II, stayed on until their retirement, and then returned home, not seeming to realize that in the meantime everyone else had left and the island had turned into a desert.

The trees had been cut for the saltworks, and without trees there was little rain and the earth dried out. The landscape is arid, with desert bushes clawing up from sandy, barren soil. Cattle and donkeys seeking the scarce shade from midday sun seem to be hiding silently behind every bush. The nights are starlit and cooled by an easterly breeze, with no sounds but the rolling sea and the occasional rustle of a wandering cow.

Adolphus Kennedy, born in 1915, lives on Salt Cay with his wife, who is three years younger. They are alone. Their four children and many grandchildren have all left. A soft-voiced man with a gentle manner, he recalls loading salt onto four-masted schooners. Mostly he remembers the weight of the sacks. The pay?

"Oh, there was no pay. They paid us nothing. The big company kept all the money."

It is easy to be romantic about the vanished Caribbean salt trade, but in truth it was similar to the history of sugarcane on other islands. Salt was built on slavery, and many thought that abolition in 1836 would mean the end of salt. But the salt merchants survived for a time because they could still get workers for near slave wages. There was no other work.

Most of the old belongers remember working for one shilling

sixpence a day, less than a dollar. "That's right," said Kennedy. "One shilling sixpence for a nine-hour day." He smiled. "But not every day. When ships were in and there was work. It was slavery," he said, revealing no bitterness in his voice.

"Of course you could buy some food for a shilling. There was rain in those days, and they could use the land. Grew corn and beans and cucumbers." Today, the land is too dry for such gardens. With no salt, little work, and less agriculture, the aged belongers live on government assistance. The Turks and Caicos is still a British colony.

The sluices are kept open and the ponds get seawater at high tides, so most are not evaporating. A few have evaporated into sand and salt crystal. In low tide, one of the canals that brings in seawater exposes two eighteenth-century cannons, weapons of war brought here to guard British salt from the Spanish, now rusting in the saltwater.

⌒

THE UNITED STATES is both the largest salt producer and the largest salt consumer. It produces over 40 million metric tons of salt a year, which earns more than $1 billion in sales revenue. The production leaders, behind the United States, in order of importance are China, Germany, Canada, and India. France has fallen to eighth place and the United Kingdom to ninth.

But little of this is table salt. In the United States, only 8 percent of salt production is for food. The largest single use of American salt, 51 percent, is for deicing roads.

American salt sources are many and varied. The Great Salt

Lake, which is the fourth largest lake in the world without an outlet, produces salt, some by Morton. Cargill operates a rock salt mine 1,200 feet below the city of Detroit. It covers more than 1,400 underground acres and has fifty miles of roads. The mine began with a disaster. In 1896, a 1,100-foot shaft was sunk but then became flooded with water and natural gas killing six and losing the investors their money. But in 1907, the mine was successfully started up.

Cargill also operates the mine on Avery Island. The McIlhenny and Avery families lease the salt mine to Cargill, the oil and gas to Exxon, and they still make the pepper sauce themselves. Paul McIlhenny, president and CEO, is the great-grandson of Edmund, who had come home from New Orleans with the seeds. He inherited Edmund's robust round features and rugged friendly eyes. "We are lucky to be in a place that not only supports agriculture but has oil, gas, and salt," he said.

The agriculture he referred to is for Tabasco sauce, which has developed into a successful international family-owned business. The peppers grown today on Avery Island are only used for seeds that are planted in Central America, where pepper picking is still cost effective. Not only does the picking require skill, since each pepper must be picked at its moment of optimum ripeness, but it is painful, backbreaking work. The powerful capsaicin can burn hands, or if the picker is careless, the face and eyes. Experiments with machine harvesting failed, and the McIlhenny family refused to experiment with chemicals that would make all the peppers ripen simultaneously. So in the 1970s, when it started to become difficult to find people in southern Louisiana willing to pick peppers, the solution was to reverse history and take seeds from Avery Island back to Mexico and Central America every year.

After one post–Civil War failure, salt mining began in earnest on Avery Island in 1898. Cargill took over in 1997. The current operation can mine nineteen tons of salt in a minute and a half and takes 2.5 million tons a year. Down in the mine more equipment is seen than actual miners. Bulldozers, tractors, jeeps, pickups, trucks, train carts, tracks, and other equipment are brought down piece by piece in a five-by-seven-by-ten-foot shaft elevator and assembled in the mine. Below, it looks like a busy nighttime construction site. A scaler, a huge machine that resembles a brontosaurus, steadily munches away at the white walls. When equipment is no longer useful, it is not considered cost effective to take it apart and bring it back up, so the mine leaves a trail of abandoned equipment, a junkyard on the side of some of the wide shafts. Salt mining has always been like that. The horses used in Wieliczka, and the mules lowered by rope underneath Detroit, never came back up either.

One of the older miners said that his father had worked fifty-two years under Avery Island, carrying salt blocks and loading them on mules. Today, the salt is trucked to crushers that break it into small enough pieces for the conveyor belts to move the salt to barges that carry it along the bayou and up the Mississippi River. A barge will hold 1,500 tons of salt.

The mine is dug in rooms called benches that are 60 feet by 100 feet with 28-foot ceilings. Once a bench is mined, a road is dug through the floor down to another level and another bench. The salt dome that is being mined is a column of solid sodium chloride, crystal clear, thought to be 40,000 feet deep—almost eight miles. The floors, the walls, the ceiling, and the uncut depths below are all between 99.25 and 99.9 percent pure. Under the miners' lamps—the first miner's safety lamp was invented by Humphry Davy—the benches appear to be black rooms. But a freshly cut bench, without the soot of machinery, is crystalline white, a room of pure salt crystal.

The vehicles are all four-wheel-drive, because the salt floor is as slippery as ice. Driving jeeps and trucks deep in the earth is like driving through a snow blizzard, at night. But it is darker than night. "It's so dark it hurts your eyes," one miner said.

The mine is currently operating at a depth of 1,600 feet, and with 38,400 feet to go, it might seem that this salt dome is an inexhaustible resource. But as the miners dig, to withstand added stress from the weight above them, the benches must be made smaller. Another problem is that salt is a good conductor of heat. The earth gets hotter closer to its center, and as they dig deeper into the earth the temperature will rise from the current ninety degrees. The heat will require more ventilation and more efficiency in machine-cooling systems. Also, the conveyor belt will get longer and longer. So the deeper they go, the more expensive the salt becomes, and salt must be cheap to be profitable. It is thought that the dome will offer another forty or fifty years of cost-effective mining, but that is a guess.

The salt is used for road deicing, industry, and pharmaceuticals. Table salt production stopped in 1982 when the energy cost of the vacuum evaporators was judged too costly. The Chinese might think that a salt dome full of oil and natural gas would have no problem with cheap energy, but in this case, the salt and the gas are operated by two separate companies that never arrived at the simple solution of ancient Sichuan.

In nearby New Iberia, a town canaled by bayous and draped in swaying moss, Avery Island salt used to be the salt of Cajun food. Ted Legnon's father was a salt worker on Avery Island, and he brought home blocks of salt for boudin sausages and cured meat. Now Ted is a butcher, and he still makes boudin, though he now makes it without hog's blood because the health department stopped the local slaughtering practices. He also uses Morton's and not local salt.

One pound of salt is used for 250 pounds of boudin, along with ground pork meat, pork liver, cooked rice, onions, bell peppers, and powdered cayenne pepper. It is all stuffed in hog's intestine and gently poached. Legnon's Butcher Shop in New Iberia sells 300 pounds of this *boudin blanc* per day, except between Christmas and New Year's, when sales rise to 500 pounds per day.

~

IN RECENT YEARS, scientists and engineers have been drawn to the ability of salt mines to preserve, because they usually have a low and steady humidity, and if not drilled too deep, an even, cool temperature. Also, salt seals. Crystals will grow over cracks. This was how the Celtic bodies had been sealed in the mine at Hallein. It is also why soy sauce makers formed a crust of salt on the top of the barrel, to make a perfect seal.

In March 1945, American troops passing through the German town of Merkers discovered a salt mine 1,200 feet underground. In it was 100 tons of gold bullion, twenty-nine rows of sacks of gold coins, and bails of international currency, including 2 million U.S. dollars. They also found more than 1,000 paintings, including Raphaels and Rembrandts. Among the booty were things of little value, such as the battered suitcases of people deported to concentration camps. The total value of the treasures, preserved in the perfect stable environment of a salt mine, was estimated at $3 billion 1945 dollars.

Because of the sealing ability of salt, it has also occurred to engineers that salt mines might be the safest place to bury nuclear waste. A Carlsbad, New Mexico, mine is being prepared for plutonium-contaminated nuclear waste that will remain toxic for the next 240,000 years. Salt will close over fractures, but how do we warn people 100,000 years from now not to open the mine? What language can be used? Suggestions include a series of grimacing masks.

The U.S. government has also stored an emergency reserve of petroleum in salt domes throughout the Gulf of Mexico area. The idea of a strategic oil reserve was first proposed in 1944. In the 1970s, it was decided to store at least 700 million barrels of oil in a select few of the 500 salt domes that have been identified in southern Louisiana and eastern Texas. But, ominous for the nuclear waste program, the domes don't always seal. The Weeks Island salt dome, not far from Avery Island, was part of the U.S. Strategic Petroleum Reserve until signs of water leakage led to fear of flaws in the dome. The oil was pumped out and the dome abandoned.

⌒

THE OWNERS OF the Dürnberg salt mine in Hallein, Austria, which has been hosting visitors since at least 1700, decided in 1989 that salt was no longer profitable and closed down the mine. But it still earns money from 220,000 visitors each year, taking them on rides on the steep, long wooden slides that were built to transport miners.

In the nineteenth century, when health spas became fashionable, many of the old brine springs saw a more lucrative alternative to making salt. In 1855, a bath was started at Salies-de-Béarn. In 1895, a red stone pseudo-Moorish palace was built to house the baths, which, in spite of continued salt production, became the town's leading economic activity. The baths are said to be beneficial for gynecological problems, rheumatism, and children with

Nineteenth-century salt making at Salies-de-Béarn. Marcel Saule

growth problems. Today, only about 750 tons of salt are made in Salies-de-Béarn every year, but this is enough to ensure that the tradition of jambon de Bayonne continues.

The part-prenants are still organized and entitled to their share of the salt production, but today, rather than being paid in large buckets of brine, they are paid in money. Each of the 564 remaining part-prenants receives approximately thirty dollars per year.

The claim that millions of years ago, algae in the brine left bromides, iodine, potassium, and other minerals, is the basis of the town business in Salsomaggiore, outside of Parma. A spa palace was built from 1913 to 1923 by architect Ugo Giusti and decorator Galileo Chini. It is considered the greatest example of Liberty architecture, the Italian version of art nouveau. Marble columns line the halls, huge staircases climb to floors of marble mosaic with wicker furniture. Murals in gold leaf on the theme of water fill the huge, high-ceilinged walls.

The brine at the spa is said to be especially helpful to those suffering from rheumatism, arthritis, and circulatory ailments. Each year, 50,000 people go to Salsomaggiore to sit in a deep turn-of-the-century tub in a tiled room and be cured in brine like a herring.

The town is a collection of 1920s and 1930s hotels and cafes, resembling a faded, out-of-fashion Riviera resort without a beach. The well-dressed clientele arrive in trickles, not waves. The spa business has been struggling of late because the Italian government has stopped covering health spas in its national health plan.

Meanwhile, the famous prosciutto di Parma are now cured with salt from Trapani. A crossbred pig has been developed that has less fat and more weight, and it is still fed on the whey from Parmigiano cheese. The pig produces a huge, round, meaty leg. The designated area for prosciutto di Parma is about forty square miles, centered on the rolling, black-soiled farmland of Langhhirano, which means "lake of frogs," originally a marsh. The popularity of these hams has turned prosciutto into a huge business, and the hundreds of thousands of hams produced every year are now cured in climatized rooms that derive no advantage from the dry winds.

⌒

FASHIONABLE PEOPLE ARE now divided into two camps. One is passionate about being healthy and eating less salt, the other is passionate about salt. The argument has been continuing since ancient times between those who think salt is healthy and those who think it is unhealthy. They both may be right. Unarguably, the body needs salt. A great deal of research indicates a relationship between high blood pressure and cardiovascular problems and eating large quantities of salt. *The Yellow Emperors Classic of Internal Medicine*, a Chinese book from the first or second century A.D., warned that salt can cause high blood pressure, which can lead to strokes. Not coincidently, one of the fatal symptoms of salt deficiency is low blood pressure. But there are also studies that refute a link between high salt intake and high blood pressure. Some studies even indicate that low-salt diets are unhealthy. The kidneys store excess sodium, and in theory someone with healthy kidneys could eat excessive salt with impunity. Sweating and urination, by design, relieve the body of salt excesses. The problem lies in the balance of sodium and potassium. But it seems that an imbalance cannot be adjusted simply by eating more or less potassium-rich vegetables versus sodium-rich salt.

The theoretical debate continues, but clinical evidence shows that people who consume large quantities of salt are not as healthy as those who don't.

Meanwhile, fashionable chefs are cooking with more salt—or more noticeable salt. It has become stylish to serve food on a bed of salt, cook it in a crust of salt, make it crunchy with lots of large crystals. More than 1,000 years ago, the Chinese were cooking in a salt crust. Chicken cooked in a crust of salt is an ancient recipe attributed to the Cantonese, though it may have originated with a south China mountain people known as the Hakka. Today, fish is cooked this way in Italy, France, Spain, and many other places. Even a fish farmer with a small restaurant by the Dead Sea in Israel cooks his fish in a salt crust. The salt seals, in the same way that cooking in clay does, but it does not make the fish or chicken salty. French chefs sometimes leave the fish unscaled to avoid salting the flesh.

In the old Guérande salt port of Le Croisic, in a 1615 thick-beamed, stone waterfront building where merchants used to buy salt for the moored ships, is a restaurant called Le Bretagne. It is one of many restaurants in the Guérande area that specializes in sea bass baked in a salt crust. It is evident from the quantity of salt used that this is a style of cooking either for the very rich or for a modern age of inexpensive salt.

BAR EN CROUTE DE SEL
(SEA BASS IN SALT CRUST)

Choose a sea bass of about two pounds for two people.

Five to seven pounds of *sel de Guérande* [gray salt from Guérande]

Five black peppercorns, thyme, rosemary, tarragon, and fennel

Clean the fish. **Do not scale it.**

Fill the stomach with the herbs and a few turns of a pepper mill.

Place the fish on a bed of coarse salt in an ovenware dish. Cover the fish with a layer of salt slightly less than an inch thick. Pat it down and moisten it with a spray humidifier.

Bake it in the oven and decorate with seaweed.—*Michèle and Pierre Coïc, Le Bretagne, Le Croisic*

AFTER THOUSANDS OF years of struggle to make salt white and of even grain, affluent people will now pay more for salts that are odd shapes and colors. In the late eighteenth century, British captain James Cook reported that the Hawaiians made excellent salt. However, he complained that on the island of Atooi, today known as Kauai, the salt was brown and dirty. The cause of this was a tradition of mixing the salt with a local volcanic red clay, *alaea*, which is brick red from a high iron content. Cook did not seem to understand that this "dirty salt" was not intended to be table salt. The salt was made for ritual blessings and religious feasts. It was also used to preserve marlin and as a medicine, especially for purification during periods of fasting. But today this dirty salt, "alaea red salt," is widely available, sought after by fine chefs and would-be gourmets.

Gray salts, black salts, salts with any visible impurities are sought out and marketed for their colors, even though the tint usually means the presence of dirt. Like the peasants in Sichuan, many consumers distrust modern factory salt. They would rather have a little mud than iodine, magnesium carbonate, calcium silicate, or other additives, some of which are merely imagined. The New Cheshire Salt Works, which does not add iodine, adds sodium hexacyanoferrate II as an anticaking agent. There is no evidence that such chemicals are harmful, and, in the case of iodine, a great deal of evidence that it is healthy. Now there is talk of adding flouride to salt for its health benefits. But modern people have seen too many chemicals and are ready to go back to eating dirt.

Then, too, many people do not like Morton's idea of making all salt the same. Uniformity was a remarkable innovation in its day, but it was so successful that today consumers seem to be excited by any salt that is different.

Among the big winners in this new salt fashion are the old-time producers from the Bay of Bourgneuf. These were the salts that Colbert had complained could be so much more sellable if only the producers would learn to make them whiter. Their impurities had always been a drawback. As recently as 1911, the French pharmacist Francis Marre, warned in his book *Défendez votre estomac contre les fraudes alimentaires*, "In general the first thing to do [when buying salt] is to make sure it is white, this would give you the best assurance that it is a pure product." The problem with the Bay of Bourgneuf salt was the dark umber clay at the bottom of the ponds. Particles got caught in the square sodium chloride crystals. But once again, Colbert was wrong. Today's consumers will pay high prices for the gray salt of Guérande, Noirmoutier, or Ile de Ré.

Many French traditions vanished in the 1980s, including kigsall, the salted pork of Guérande. In Noirmoutier, the salt business almost died between 1986 and 1994, with only twenty-one salt makers left. In 1995, a group of locals and off-islanders formed a salt cooperative to bring back traditional salt making, and now almost 160 work on the island's salt ponds and market their salt through the cooperative. Guérande, Noirmoutier, and Ile de Ré have all attracted French people longing to return to

A nineteenth-century postcard of salt rakers harvesting in the Guérande region. Collection of Gildas Buron, Musée des Marais Salants, Batz-sur-Mer

agriculture. One-third of the salt makers on Noirmoutier are younger than thirty-five.

The paludiers of Guérande used to be locals in that part of Brittany who passed on the work from father to son. But today only 20 percent of the 300 paludiers are local. After two genera-tions of French leaving their villages and their agricultural way of life and moving to the cities, there is a significant minority doing the reverse. They leave Paris to raise ducks in Périgord or oysters on the Atlantic. And some come to Brittany to rake salt in a way that is so traditional that fiberglass poles and rubber tires for wheelbarrows are the only discernible changes in technology since before the Revolution.

Unlike with the big companies, here the future is quality, not quantity. They command high prices for their salt because it is a product that is handmade and traditional in a world increasingly hungry for a sense of artisans. They make two kinds of salt, the gray salt and fleur de sel. The light, brittle fleur de sel crystals are ten times more expensive than costly gray salt. French salt mak-ers do battle in court over what is true fleur de sel. Guérande has sued Aigues-Mortes—all the more suspect since it was bought by "the Americans"—over its use of the term.

But fleur de sel is not unique to Brittany, and may be as old as making sea salt. In the second century B.C., Cato gave instruc-tions for making fleur de sel in *De agricultura:*

Fill a broken-necked amphora with clean water, place in the sun. Suspend in it a strainer of ordinary salt. Agitate and refill repeatedly; do this several times a day until salt remains two days undissolved. A test: drop in a dried anchovy or an egg. If it floats, the brine is suitable for steeping meat, cheese or fish for salting. Put out this brine in pans or baking dishes in the sun, and leave in the sun until crystallized. This gives you "flower of salt." When the sky is cloudy, and at night, put indoors; put in the sun daily when the sun shines.

Skimming the ponds at Guérande for fleur de sel as the sun sets, paludiers can look up and see the silhouette of the Moorish stone steeple of Saint-Guénolé in Batz-sur-Mer across the grassy wetlands. Near the church is the distinct smell of butter and a long line of people in front of the bakery, Biscuits Saint-Guénolé. The bakery was started in the 1920s with the recipes of a woman who sold cakes in the neighborhood. Breton baking is about salt and butter. Although modern refrigeration has made unsalted butter easily available, Bretons insist that salt brings out the flavor of butter.

Kouing amann, the name of one of the most famous cakes of the region, is Celtic for "piece of butter." Gérard Jadeau, who runs the shop, says that in his kouing amann, butter makes up half of the total ingredients. It is a buttery dough, layered like puff pastry, rolled flat. Butter is spread on it. Then it is rolled and sliced, and the slices are arranged in a baking dish and put in the oven.

It is not easy to get this much butter in a cake. The trick is a moderate oven. Too hot, and the butter will separate; too cool, and the butter will keep the dough from baking. But even with this quantity of butter, Breton bakers will say that what gives the cake its buttery flavor is the saltiness.

Jadeau, between ovens, wrote down on the back of his business card the following recipe for his popular butter cookies:

GALETTE FINE

55 kg flour
30 kg sugar
20 kg butter

        8 kg eggs
        1k 200 salt
    Mix the dough, let it rest a half an hour, fashion it into
cookies and bake.

"The trick," he said teasingly, "is always mixing in the right order." But he refused to say what the correct order was. Clearly,
the flour is last, because in baking, flour is always the last ingredient. The butter is probably mixed with the salt first. Jadeau salts
his own butter with salt from the same Batz-sur-Mer producer of
Guérande salt that his shop has been using since 1920.
    Does he use the gray salt?
    "No."
    Fleur de sel?
    "No, that's too expensive."
    He gets a Batz producer to wash the gray salt until it is white
and then crush it fine. "I don't think the gray salt is clean," he
said. "It has dirt in it. This salt I use you could get anywhere. But
I am here so I get it here."
    In the past, this kind of fine, white salt was called in Celtic
holen gwenn, white salt, and was rare and expensive, only for the
best tables and the finest salted foods. The gray was the cheap
everyday salt. The relative value of the white and gray salt is a
question of supply, demand, and labor, but also culture, history,
and the fashion of the times.
    Why should salt that is washed be cheaper than salt with dirt?
Fixing the true value of salt, one of earth's most accessible commodities, has never been easy.

Acknowledgments

⌒

ONE OF THE remarkable experiences of journalism is that as you travel from place to place, people take the time to help you. I especially want to thank Shane Bernard for his research and helpfulness at Avery Island, Gildas Buron for his thoughtful assistance at the Musée Intercommunal des Marais Salants in Batz-sur-Mer, Stephen Fawkes for his help in Great Inagua, Andreu Galera for insights and guidance in Cardona, Cheng Jinfeng for his generous and good-humored assistance in Zigong, Antonio d'Ali for his hospitality in his Trapani saltworks, Marzio Dall'Acqua for his help at the archives in Parma, Professor Guo Zhengzhong of Beijing for sharing his vast knowledge of salt history and his beautiful calligraphy, Oded Harel for his help at the Dead Sea, Bo Masser and family for their hospitality and enthusiastic assistance in Sweden, Paul McIlhenny for his hospitality at Avery Island, Roy Moxham for giving me an early look at his wonderful book, *The Great Hedge of India*, Fred Plotkin for his help with Italian, Marcel Saule for his generous help in Salies-de-Béarn, Bryan Sheedy for his passionate look at Salt Cay, Peter Sherratt for his assistance in Cheshire, Jill Singleton for her help in San Francisco Bay, Laura Trombetta for her Chinese translation, the generous sharing of her vast knowledge, and her sense of fun and relentless curiosity as a travel companion, Marianne Vleeschhouwer and Anne Kupfer for their help with Dutch.

An overdue thank you to two institutions that make food history research possible in New York: to Nach Waxman and his magical bookstore, Kitchen Arts & Letters; and to the New York Public Library, whose staff has been helping me all of my reading life in that elegant stone-and-polished wood palace with underground vaults full of everything, including more than 1,000 books on salt.

A special thanks to my editor, Nancy Miller, for her craft and determination to make me my best, and to my agent, Charlotte Sheedy, for her good humor and wisdom. To George Gibson for his friendship and perfect publishing, and to Linda Johns for her belief in this and for not allowing me to talk myself out of it. Also thanks to Sarah Walker and Sasha Yazdgerdi for all their help. A special thanks to my brother Paul Kurlansky for his insights into chemistry, and to my brother Steven Kurlansky and sister Ellen Brown for their loving support. And a most special thanks to my beautiful wife, Marian Mass, for making me laugh, for making me think, and for her wondrous smile, bright enough to light planets.

Bibliography

BOOKS

Salt History

Adshead, S. A. M. *Salt and Civilization*. New York: St. Martin's Press, 1992.

Aggarwal, S. C. *The Salt Industry in India*. New Delhi: Government of India Press, 1976.

Andrews, Anthony P. *Maya Salt Production and Trade*. Tucson: University of Arizona Press, 1983.

Bach, Antoni. *Història de Cardona*. Barcelona: Curial Documents de Cultura, 1992.

Badia, Enric. *La sal, support d'uns pobles*. Manresa, Spain: Angle Editorial, 1996.

Baudry-Weulersee, Delphine. *Yantie lun: Dispute sur le sel et le fer* (Discourse on salt and iron). Trans. Jean Levi and Pierre Baudry. Paris: J. Lansmann and Seghers, 1978.

Bergier, Jean-François. *Une histoire du sel*. Fribourg, Switzerland: Office du Livre, 1982.

Boudet, Gérard. *La Renaissance des Salins du Midi de la France au XIXe Siècle.* Paris: Salins du Midi, 1995.

Bravo Dueñas, Miguel. *Mina de Cardona (1929–1989).* Grata Lecura, 1998.

Bridbury, A. R. *England and the Salt Trade in the Later Middle Ages.* London: Oxford University Press, 1955.

Brownrigg, William. *The Art of Making Common Salt.* London, 1748.

Buron, Gildas. *Bretagne des Marais Salants: 2000 ans d'histoire.* Morlaix: Skol Ureizh, 1999.

Calvert, Albert F. *A History of the Salt Union.* London: Effingham Wilson, 1913.

———. *Salt in Cheshire.* New York: Spon and Chamberlain, 1915.

———. *Salt and the Salt Industry.* London: Sir Isaac Pitman and Sons, 1919.

Changyong, Zhong, Huang Jian, and Lin Jianyu. *A Famous City Zigong: Zigong Salt China* (in Chinese) Zigong: Sichuan People's Publishing House, 1993.

Choudhury, Sadananda. *The Economic History of Colonialism: A History of British Salt Policy in Orissa.* Delhi: Inter-India Publications, 1979.

Cole, L. Heber. *Report on the Salt Deposits of Canada and the Salt Industry.* Ottawa: Government Printing Bureau, 1915.

Collins, John. *Salt and Fishery, Discourse Thereof.* London, 1682.

De Person, Françoise. *Bateliers contrebandiers du sel.* Rennes: Editions Ouest-France, 1999.

Dopsch, Heinz, Barbara Heuberger, and Kurt W. Zeller, ed. *Salz.* Salzburg: Salzburger Landesausstellung, 1994.

Dunoyer de Segonzac, Gilbert. *Les chemins du sel.* Paris: Gallimard, 1991.

Eskew, Garnett Laidlaw. *Salt: The Fifth Element.* Chicago: J. G. Ferguson and Associates, 1948.

Figuier, Louis. *L'industrie du sel.* Paris, 1876.

Hocquet, Jean-Claude. *Le sel de la terre.* Paris: Editions du May, 1989.

———. *Le sel et la fortune de Venise.* Vol. 1, *Production et monopole.* Vol. 2, *Voiliers et commerce en Méditerranée, 1200–1650.* Lille: Publications de L'Université de Lille III, 1979.

Kaufman, Dale W., ed. *Sodium Chloride: The Production and Properties of Salt and Brine.* New York: Hafner Publishing, 1968.

Labarthe, Jean. *Salies-de-Béarn historique et anecdotique*. Salies: Les Amis du Vieux Salies, 1996.

Le Foll, Nathalie. *Le sel*. Paris: Editions du Chêne, 1997.

Le Roux, Pierre, and Jacques Ivanoff, eds. *Le sel se la vie en Asie du Sud-est*. Bangkok: Prince of Songkla University, 1993.

Lemonnier, Pierre. *Paludiers de Guérande: Production du sel et histoire économique*. Paris: Institut d'Ethnologie-Musée de l'Homme, 1984.

Lonn, Ella. *Salt as a Factor in the Confederacy*. New York: Walter Neale, 1933.

Lovejoy, Paul E. *Salt of the Desert Sun: A History of Salt Production and Trade in the Central Sudan*. Cambridge: Cambridge University Press, 1986.

Marlan, Stanton, ed. *Salt and the Alchemical Soul: Three Essays by Ernest Jones, C. G. Jung, and James Hillman*. Woodstock, Conn.: Spring Publications, 1995.

Mendizabal, Miguel Othan de. *Influencia de la sal en la distribución geografica de los grupos indigenas de México*. Mexico City: Museo Nacional de Arqueologia, Historia y Etnografia, 1928.

Multhauf, Robert P. *Neptune's Gift: A History of Common Salt*. Baltimore: Johns Hopkins University Press, 1996.

Nenquin, Jacques. *Salt: A Study in Economic Prehistory*. Bruges: De Tempel, 1961.

Quinn, William P. *The Saltworks of Historic Cape Cod*. Orleans: Parnassus Imprints, 1993.

Piégay, Joseph. *Les mulets de sel: Une rebellion paysanne dans le pays du Verdun en 1710*. Mane, Haute-Provence: Les Alpes de Lumière, 1998.

Stealey, John E. *The Antebellum Kanawha Salt Business and Western Markets*. Lexington: University Press of Kentucky, 1993.

Weiss, Harry B., and Grace M. Weiss. *The Revolutionary Saltworks of the New Jersey Coast*. Trenton: Past Times Press, 1959.

Food and Food History

Anonymous. *The New Family Receipt-Book: Containing 700 Truly Valuable Receipts in Various Branches of Domestic Economy Selected from the Works of British and Foreign Writers of Unquestionable Experience and Au-*

thority, and from the Attested Communications of Scientific Friends. London: John Murray, 1810.

Allen, Brigid. *Food: An Oxford Anthology*. Oxford: Oxford University Press, 1994.

Anderson, E. N. *The Food of China*. New Haven: Yale University Press, 1988.

Anthimus. *De obseruatione ciborum* (On the observance of food). Trans. Mark Grant. Devon: Prospect Books, 1996.

Apicius. *Cookery and Dining in Imperial Rome*. Ed. and trans. Joseph Dommers Vehling. New York: Dover, 1977.

Archestratus. *The Life of Luxury*. Trans. John Wilkins and Shaun Hill. Devon: Prospect Books, 1994.

Artusi, Pellegrino. *La scienza in cucina e l'arte di mangiar bene*. 1891. Reprint, San Casciano: Sperling and Kupfer Editori, 1991.

Beeton, Isabella. *Beeton's Book of Household Management*. London: S. O. Beeton, 1861. Reprint, London: Jonathan Cape, 1968.

Billings, John D. *Hardtack and Coffee*. Boston: George M. Smith, 1887.

Bober, Phyllis Pray. *Art, Culture, and Cuisine: Ancient and Medieval Gastronomy*. Chicago: University of Chicago Press, 1999.

Boni, Ada. *Il talismano della felicità*. Rome: Colombo, 1997.

Brécourt-Villars, Claudine. *Mots de table, mots de bouche*. Paris: Editions Stock, 1996.

Brereton, Georgina E., and Janet M. Ferrier, ed. *Le mèsnagier de Paris*. Paris: Le Livre de Poche, 1994.

Brillat-Savarin, Jean Anthelme. *Physiologie du goût*. Paris: Flammarion, 1982.

Brothwell, Don, and Patricia Brothwell. *Food in Antiquity: A Survey of the Diet of Early Peoples*. Baltimore: Johns Hopkins University Press, 1969.

Bullock, Helen. *The Williamsburg Art of Cookery, or Accomplish'd Gentlewoman's Companion: Being a Collection of Upwards of Five Hundred of the Most Ancient and Approv'd Recipes in Virginia Cookery*. Williamsburg: Colonial Williamsburg, 1965.

Buhez Association. *Quand les Bretons passent à table*. Rennes: Editions Apogée, 1994.

Capel, José Carlos. *Manual del pescado*. San Sebastián: R and B, 1995.

Cato. *De agricultura* (On farming). Trans. Andrew Dalby. Devon: Prospect Books, 1998.

Correnti, Pino. *Il libro d'oro della cucina e dei vini di Sicilia*. Milano: Mursia, 1976.

Cost, Bruce. *Bruce Cost's Asian Ingredients: Buying and Cooking the Staple Foods of China, Japan, and Southeast Asia*. New York: William Morrow, 1988.

Couderc, Philippe. *Les plats qui ont fait La France: De l'andouillette au vol-au-vent*. Paris: Julliard, 1995.

Curtis, Robert I. *Garum and Salsamenta: Production and Commerce in Materia Medica*. Leiden: E. J. Brill, 1991.

Da Silva, Elian, and Dominique Laurens. *Fleurines & Roquefort*. Rodez: Editions du Rouergue, 1995.

Dalby, Andrew. *Siren Feasts: A History of Food and Gastronomy in Greece*. London: Routledge,1996.

Dalby, Andrew, and Sally Grainger. *The Classical Cookbook*. Los Angeles: J. Paul Getty Museum, 1996.

David, Elizabeth. *English Bread and Yeast Cookery*. 1997. Reprint, Newton, Mass.: Biscuit Books, 1994.

―――. *Harvest of the Cold Months: The Social History of Ice and Ices*. London: Viking, 1994.

Davidson, Alan. *North Atlantic Seafood*. London: Macmillan, 1979.

―――. *The Oxford Companion to Food*. Oxford: Oxford University Press, 1999.

de la Falaise, Maxime. *Seven Centuries of English Cooking: A Collection of Recipes*. Ed. Arabella Boxer. New York: Grove Press, 1973.

Drummond, J. C., and Anne Wilbraham. *The Englishman's Food*. London: Pimlico, 1994.

Dumas, Alexandre. *Le grand dictionnaire de cuisine*. Vol. 3, Poissons. 1873. Reprint, Payré: Edit-France, 1995.

Eaton, Mary. *The Cook and Housekeepers Complete and Universal Dictionary*. Bungay, England: 1822.

Evelyn, John. *Acetaria: A Discourse of Sallets*. 1699. Reprint, Devon: Prospect Books, 1996.

Faccioli, Emilio, ed. *L'arte della cucina in Italia*. Torino: Giulio Einaudi, 1987.

Fitzgibbon, Theodora. *A Taste of Ireland*. London: Weidenfeld and Nicolson, 1968.

Flandrin, Jean-Louis, and Massimo Montanari. *Histoire de L'Alimentation*. Paris: Librarie Arthème Fayard, 1996.

Glasse, Hannah. *The Art of Cookery Made Plain and Easy Which Far Exceeds Any Thing of the Kind Ever Yet Published by a Lady*. London: printed by the author, 1747.

(the widow) Gontier. *Traité de l'olivier, l'histoire et la culture de cet arbre, les différentes manieres d'exprimer l'huile d'olive, celles de la conserver, &c*. 1784. Reprint, Nîmes: C. Lacour, 1990.

Gozzini Giacosa, Ilaria. *A Taste of Ancient Rome*. Trans. Anna Herklotz. Chicago: University of Chicago Press, 1992.

Grimod de La Reynière, Alexandre. *Ecrits gastronomiques*. "Almanach des gourmands," 1803, and Manuel des Amphiryions, 1808. Reprint, Paris: Bibliothèques 10/18, 1997.

Hagen, Ann. *A Second Handbook of Anglo-Saxon Food and Drink: Production and Distribution*. Norfolk, England: Anglo-Saxon Books, 1995.

Hale, Sarah Josepha. *The Good Housekeeper, or The Way to Live Well and to Be Well while We Live*. Boston: Otis, Broader, 1841.

Henisch, Bridget Ann. *Fast and Feast: Food in Medieval Society*. University Park: Pennsylvania State University Press, 1990.

Hieatt, Constance B., ed. *An Ordinance of Pottage: An Edition of the Fifteenth Century Culinary Recipes in Yale University's Ms Beinecke 163*. London: Prospect, 1988.

Hope, Annette. *A Caledonian Feast*. London: Grafton, 1989.

———. *Londoners' Larder*. Edinburgh: Mainstream Publishing, 1990.

Hume, Audrey Noel. *Food*. Williamsburg, Va.: Colonial Williamsburg Foundation, 1990.

Innis, Harold A. *The Cod Fisheries: The History of an International Economy*. New Haven: Yale University Press, 1940.

Jarvis, Norman R. *Curing Fishery Products*. Kingston, Mass.: Teaparty Books, 1987.

Kasper, Lynne Rossetto. *The Splendid Table: Recipes from Emilia-Romagna, the Heartland of Northern Italian Food*. New York, William Morrow, 1992.

Ky, Tran, and François Drouard. *Le chou et la choucroute*. Condé-sur-Noireau: Editions Charles Corlet, 1997.

Laudan, Rachel. *The Food of Paradise: Exploring Hawaii's Culinary Heritage*. Honolulu: University of Hawaii Press, 1996.

Launay, André. *Caviare and After: The Truth About Luxury Foods*. London: Macdonald, 1964.

Levy, Esther. *Jewish Cookery Book*. Philadelphia, 1871. Reprint, Cambridge: Applewood Books, 1988.

Marre, Francis. *Défendez votre estomac contre les fraudes alimentaires*. Paris: H. Malet, 1911.

May, Robert. *The Accomplisht Cook*. 1685. Reprint, Devon: Prospect Books, 1994.

McClane, A. J. *The Encyclopedia of Fish Cookery*. New York, Henry Holt, 1977.

McGee, Harold. *On Food and Cooking: The Science and Lore of the Kitchen*. New York: Scribner's, 1984.

Molokhovets, Elena, *A Gift to Young Housewives*. Trans. Joyce Toomre. 1897. Reprint, Bloomington: Indiana University Press, 1998.

Montagné, Prosper. *Larousse gastronomique*. Paris: Larousse, 1938.

Montefiore, Judith. *The Jewish Manual*. London. T. and W. Boone, 1846.

Moore, John Hammond, comp. and ed. *The Confederate Housewife: Receipts and Remedies, Together with Sundry Suggestions for Garden, Farm, and Plantation*. Columbia, S.C.: Summerhouse Press, 1997.

Peachey, Stuart. *Civil War and Salt Fish: Military and Civilian Diet in the Seventeenth Century*. Leigh on Sea, Essex: Partizan Press, 1988.

Peterson, T. Sarah. *Acquired Taste: The French Origins of Modern Cooking*. Ithaca: Cornell University Press, 1994.

Platina. *De honesta voluptate et valetudine* (On right pleasure and good health). Trans. Mary Ella Milham. Tempe, Ariz.: Medieval and Renaissance Texts and Studies, 1998.

Prato, Caterina. *Manuale di cucina per principianti e per cuoche già pratiche*. Milan: Anonima Libraria Italiana, 1923.

Reboul, J.-B. *La cuisinière Provençal*. Marseilles: Tacussel, 1910.

Reynaud, Joseph. *Guide pratique de la culture de l'olivier: Son fruit et son huile*. Nîmes: C. Lacour, 1998.

Riddervold, Astri, and Andreas Ropeid, ed. *Food Conservation*. London: Prospect Books, 1988.

Root, Waverley. *Food*. New York: Simon and Schuster, 1980.

Root, Waverley, and Richard de Rochemont. *Eating in America: A History*. New York: William Morrow, 1976.

Schoeneck, Annelies. *Making Sauerkraut and Pickled Vegetables at Home: The Original Lactic Acid Fermentation Method*. Vancouver, British Columbia: Alive Books, 1988.

Schweid, Richard. *Hot Peppers: Cajuns and Capsicum in New Iberia, Louisiana*. Berkeley, Calif.: Ten Speed Press, 1989.

Scully, Terence, ed. *The Neapolitan Recipe Collection* (Cuoco Napoletano). Ann Arbor: University of Michigan Press, 2000.

———. *The Viandier of Taillevent*. Ottowa: University of Ottowa Press, 1988.

Shactman, Tom. *Absolute Zero and the Conquest of Cold*. Boston: Houghton Mifflin, 1999.

Simeti, Mary Taylor. *Pomp and Sustenance: Twenty-five Centuries of Sicilian Food*. Hopewell, N.J.: Ecco Press, 1988.

Simmons, Amelia. *American Cookery, or The Art of Dressing Viands, Fish, Poultry and Vegetables, and the Best Modes of Making Pastes, Puffs, Pies, Tarts, Puddings, Custards, and Preserves, and All Kinds of Cakes, from the Imperial Plumb to Plain Cake. Adapted to This Country and All Grades of Life*. Hartford, Conn.: published according to an act of Congress, 1796.

Simon, André L. *A Concise Encyclopedia of Gastronomy*. London: Collins, 1952.

Simoons, Frederick J. *Food in China: A Cultural and Historical Inquiry*. Boca Raton: CRC Press, 1991.

Sloat, Caroline, ed. *Old Sturbridge Village Cookbook*. Old Saybrook, Conn: Globe Pequot Press, 1984.

Smith, Andrew F. *Pure Ketchup: A History of America's National Condiment*. Columbia: University of South Carolina Press, 1996.

Smith, Eliza. *The Compleat Housewife*. London, 1758.

Spaulding, Lily May, and John Spaulding. *Civil War Recipes: Receipts from the Pages of Godey's Lady's Book*. Lexington, Ky.: University Press of Kentucky, 1999.

Tannahill, Reay. *Food in History*. New York: Stein and Day, 1973.

Tante Marie. *La véritable cuisine de famille*. Paris: A. Taride, 1926.

Thibaut-Comelade, Elaine. *La table médiévale des Catalans*. Perpignan: Les Presses du Languedoc, 1995.

Thorne, Stuart. *The History of Food Preservation*. Totowa, N.J.: Barnes and Noble Books, 1986.

Toussaint-Samat, Maguelonne. *History of Food*. Trans. Anthea Bell. Oxford: Blackwell, 1992.

Trager, James. *The Food Chronology: A Food Lover's Compendium of Events and Anecdotes, from Prehistory to the Present*. New York: Henry Holt, 1995.

Tsuji, Shizuo. *Japanese Cooking: A Simple Art*. Tokyo: Kodansha, 1980.

Visser, Margaret. *Much Depends on Dinner: The Extraordinary History and Mythology, Allure and Obsessions, Perils and Taboos of an Ordinary Meal*. New York: Macmillan, 1986.

Wheaton, Barbara Ketcham. *Savoring the Past: The French Kitchen and Table from 1300 to 1789*. New York: Scribner, 1996.

Wilkins, John, David Harvey, and Mike Dobson, eds. *Food in Antiquity*. Exeter: University of Exeter Press, 1996.

Williams, R. Omosunlola. *Miss William's Cookery Book*. London: Longman Green, 1957.

Wilson, C. Anne. *Food and Drink in Britain: From the Stone Age to the Nineteenth Century*. Chicago: Academy Publishers, 1991.

Wright, Clifford A. *A Mediterranean Feast*. New York: William Morrow, 1999.

Science and Science History

Cobb, Cathy, and Harold Goldwhite. *Creations of Fire: Chemistry's Lively History from Alchemy to the Atomic Age*. New York: Plenum Press, 1995.

Day, David T. *Eighteenth Annual Report of the United States Geological Survey to the Secretary of the Interior, 1896–97*, Washington: Government Printing Office, 1897.

Halbouty, Michel T. *Salt Domes: Gulf Region, United States, and Mexico*. Houston: Gulf Publishing, 1967.

Osborne, Roger. *The Floating Egg: Episodes in the Making of Geology*. London: Pimlico, 1998.

General History

Barber, Elizabeth Wayland. *The Mummies of Ürümchi*. New York: W. W. Norton, 1999.

Bourne, E. G. *Narratives of the Career of Hernando De Soto*. Vol. I. Ed. E. G. Bourne. New York, 1904.

Boutin, Emile. *La Baie de Bretagne et sa contrebande*. Laval: Siloë, 1993.

Chadha, Yogesh. *Gandhi: A Life*. New York: John Wiley and Sons, 1997.

Chalmers, Harvey. *The Birth of the Erie Canal*. New York: Bookman Associates, 1960.

Condor, George E. *Stars in the Water: The Story of the Erie Canal*. Garden City, N.Y.: Doubleday, 1974.

Cunliffe, Barry. *The Ancient Celts*. London: Penguin, 1997.

Díaz del Castillo, Bernal. *Verdadera historia de la conquista de la Nueva España*. Mexico City: Ediciones Mexicanas, 1950.

Ebrey, Patricia Buckley. *China: Cambridge Illustrated History*. Cambridge: Cambridge University Press, 1996.

Fischer, Louis. *The Life of Mahatma Gandhi*. New York: Harper and Row, 1983.

Herm, Gerhard. *The Celts: The People Who Came out of the Darkness*. New York: Barnes and Noble, 1993.

Herodotus. *The Persian Wars*. Bks. 1 and 2.. Trans. A. D. Godley. Cambridge: Harvard University Press, 1999.

Ibn Battuta. *The Travels of Ibn Battuta*. Vol 4. Trans. C. Defrémery and B. R. Sanguinetti. London: Hakluyt Society, 1994.

Ikram, Salima, and Aidan Dodson. *The Mummy in Ancient Egypt: Equipping the Dead for Eternity*. London: Thames and Hudson, 1998.

July, Robert W. *A History of the African People*. New York: Charles Scribner and Son, 1974.

Kupperman, Karen Ordahl, ed. *Captain John Smith: A Select Edition of His Writings*. Chapel Hill: University of North Carolina Press, 1988.

Lane, Frederic C. *Venice: A Maritime Republic*. Baltimore: Johns Hopkins University Press, 1973.

Long, E. B., with Barbara Long. *The Civil War Day by Day: An Almanac, 1861–1865*. Garden City, N.Y.: Doubleday, 1971.

Magnus, Olaus. *A Description of the Northern Peoples*. Trans. Peter Fisher. 1555. Reprint, London: Hakluyt Society, 1996.

Moxham, Roy. *The Great Hedge of India*. London: Constable, 2001.

Needham, Joseph. *Science and Civilization in China*. Vol. 4, *Physics and the Physical Technology*. Vol. 5, *Chemistry and Chemical Technology*. Cambridge: Cambridge University Press, 1954.

Pagden, Anthony, trans. and ed. *Cortés, Hernan, Letters from Mexico*. New Haven: Yale University Press, 1986.

Palissy, Bernard. *Le moyen de devenir riche, et la manière véritable par laquelle tous les hommes de la France pourront apprendre a multiplier et augmenter leurs tresors et possessions*. Paris: Robert Foüet, 1636.

Paludan, Ann. *The Chinese Spirit Road: The Classical Tradition of Stone Tomb Statuary*. New Haven: Yale University Press, 1991.

———. *Chronicle of the Chinese Emperors: The Reign by Reign Record of the Rulers of China*. London: Thames and Hudson, 1998.

Pliny the Elder. *Natural History: A Selection*. Trans. John F. Healy. New York: Penguin Books, 1991.

Polo, Marco. *The Travels of Marco Polo*. Trans. Ronald Latham. New York: Penguin Books, 1958.

Powell, T. G. E. *The Celts*. London: Thames and Hudson, 1980.

Prescott, William. *History of the Conquest of Mexico*. 1819.

Roberts, J. A. G. *A Concise History of China*. Cambridge: Harvard University Press, 1999.

Sadler, H. E. *Turks Islands Landfall: A History of the Turks and Caicos Islands*. Grand Turk: 1997.

Sahagún, Fray Bernardino de. *Historia general de las cosas de la Nueva España*. Mexico City: Porrua, 1969.

Schama, Simon. *Citizens: A Chronicle of the French Revolution*. New York: Alfred A. Knopf, 1989

Smith, Adam. *An Inquiry into the Nature and Causes of the Wealth of Nations*. Ed. Edwin Cannan. 1776. Reprint, New York: Modern Library, 1937.

Tacitus. *The Agricola and the Germania*. Trans. H. Mattingly. New York: Penguin Books, 1970.

Testart, Alain. *Des mythes et des croyances: Esquisse d'une théorie générale*. Paris: Editions de la Maison des Sciences de l'homme, 1991.

Thoreau, Henry David. *Cape Cod*. New York: Bramhal House, 1951.

DOCUMENTS

L'Assemblée Nationale. "Décret de L'Assemblée Nationale du 22 Mars 1790 qui annulle les Procés criminel des Gabelles; et qui accorde la grace et la liberté de ceux qui font détenu."

Avery, Mary Eliza, Cookbook, McIlhenny Company and Avery Island, Inc. archives; Avery Island, Louisiana.

Benton, Thomas Hart. "Speech of Mr. Benton of Missouri on the Repeal of the Salt Tax." Washington, April 22, 1840.

Hunter, Helen Virginia. "The Ethnography of Salt in Aboriginal North America". Ph.D. diss., University of Pennsylvania, 1940.

Lowndes, Thomas. *Brine-salt Improved, or The Method of Making Salt from Brine, That Shall Be as Good or Better Than French Bay-salt.* London, 1746.

Morton Salt Company. "The History of Salt" (pamphlet). 1956.

XXXIX Assemblea Intercomarcal D'Estudiosos. Vols. I and II. Cardona, 1994.

Tobacco and Salt Industries in Japan. Tokyo: Japan Monopoly Corporation, 1963.

ARTICLES

Dugger, Celia W. "Gandhi's Spirit Hovers as India Debates Iodized Salt." *New York Times*, November 2, 2000.

"A Foreigner Makes Millions for China. An Interview with Sir Richard Dane on the Reformed Salt Gabelle." *Asia* (New York) (July 1917).

Galera i Pedrosa, Andreu. "La sal de Cardona." *Dovella: Revista cultural de la Catalunya Central*, October 1994.

Haines, Charles G. "Considerations on the Great Western Canal, from the Hudson to Lake Erie: With a View of Its Expense, Advantages, and Progress." Brooklyn: Spooner and Worthington, 1818.

Hosis, Sir Alexander. "The Salt Production and Salt Revenue of China." *Nineteenth Century and After* (London) 75, no. 446 (April 1914).

Kremmer, Christopher. "With a Handful of Salt." *Boston Globe*, November 28, 1999.

Onishi, Norimitsu,"In Sahara Salt Mine, Life's Not So Grim." *New York Times*, February 13, 2001.

————. "Where Dwelling Is Kept from Dune, One Scoop at a Time." *New York Times*, January 14, 2000.

Rabinowitz, Alan. "The Price of Salt." *Natural History* 109, No. 7 (September 2000).

Rochester, Mary. "The Northwich Poor Law Union and Workhouse." Cheshire: Cheshire Libraries and Museums.

————. "Rock Salt Mining in Cheshire." Cheshire: Cheshire Libraries and Museums.

————. "Salt & Subsidence." Cheshire: Cheshire Libraries and Museums.

————. "Working Conditions in the Cheshire Salt Industry." Cheshire: Cheshire Libraries and Museums.

Rushdie, Salman. *Time*, April 13, 1998.

Rutherford, John. "Facts and Observations in Relation to the Origin and Completion of the Erie Canal." New York: N. B. Holmes, 1825.

Saule, Marcel. *Le rituel de la distribution de l'eau salée à Salies sous l'ancien régime*.

Schouten, Diny. "Groninger Vis." *Vrij Nederland* no. 18, May 8, 1999.

Squire, Mary. "Social and Environmental Conditions in the Salt Industry." Cheshire: Cheshire Libraries and Museums.

West, Donald. "The Cheshire Salt Industry in Tudor and Stuart Times (1485–1714)." Cheshire: Cheshire Libraries and Museums.

Whitby, Jonathan E. "The Sturgeon Fishers of Russia." *Wide World Magazine*, April 1906.

Index